Peasants Negotiating a Global Policy Space

Being the public voice of over 180 member organisations across nearly 90 countries, La Vía Campesina, the global peasant movement, has planted itself firmly on the international scene. This book explores the internationalisation of the movement, with a specific focus on the engagement of peasants in the processes of the Committee on World Food Security (CFS).

With a focus on agency (the capacity to act), this book explores the opportunities and challenges for mobilised peasants to engage directly in the global policy processes within the CFS. Since the reform of the CFS in 2009, civil society actors have engaged in the policy processes of this UN Committee from a self-designed and autonomous global Civil Society Mechanism. Rather than analysing La Vía Campesina's relationship with the CFS from a 'success-failure' viewpoint, the book examines the strategies, tensions, debates, and reconfigurations arising from rural actors engaging in a formal UN Committee. The book sheds light on the interaction between social movement activists seeking to become actors (through participation and voice) and the conditions necessary for this to happen.

Whereas the expectation of the dominant literature is that social movements will either disappear or institutionalise, the book presents empirical evidence that La Vía Campesina is building a much more sophisticated model. The direct participation of representatives of peasant organisations in the CFS processes is highlighted as a pioneering example of a more complex, inclusive and democratic foundation for global policy-making.

Ingeborg Gaarde holds a PhD in Sociology from the Ecole des Hautes Etudes en Sciences Sociales (EHESS), Paris, France. She has more than ten years of experience working with civil society organisations on five continents. She has divided her time between research and consulting for international NGOs and the UN, mainly on issues related to natural resources conflicts, tenure governance and people's participation in global policy-making processes.

"This book provides an important perspective on social movement advocacy for alternative agrarian futures."

– *Philip McMichael, Cornell University, USA*

"This book is a much-needed analytical perspective that goes beyond classic dichotomies … and explores how actors seek to overcome these dilemmas by fostering complementarities between different forms of political activism."

– *Geoffrey Pleyers, Université Catholique de Louvain, Belgium*

"La Vía Campesina peasants show how rural actors could not be more modern, more both local and global. They talk like the postmodern actors in the name of dignity and democracy and their space is planetary. A beautiful lesson of sociology."

– *Michel Wievioka, Ecole des Hautes Etudes en Sciences Sociales (EHESS), Paris, France*

Routledge Studies in Food, Society and the Environment

Civic Engagement in Food System Governance
A comparative perspective of American and British local food movements
Alan R. Hunt

Biological Economies
Experimentation and the politics of agri-food frontiers
Edited by Richard Le Heron, Hugh Campbell, Nick Lewis and Michael Carolan

Food Systems Governance
Challenges for justice, equality and human rights
Edited by Amanda L. Kennedy and Jonathan Liljeblad

Food Literacy
Key concepts for health and education
Edited by Helen Vidgen

Sustainable Urban Agriculture and Food Planning
Edited by Rob Rogemma

Transforming Gender and Food Security in the Global South
Edited by Jemimah Njuki, John R. Parkins and Amy Kaler

Urban Food Planning
Seeds of Transition in the Global North
Rositsa T. Ilieva

Food Consumption in the City
Practices and Patterns in Urban Asia and the Pacific
Edited by Marlyne Sahakian, Czarina Saloma and Suren Erkman

Eating Traditional Food
Politics, Identity and Practices
Edited by Brigitte Sébastia

Feeding Cities
Improving local food access, security and sovereignty
Edited by Christopher Bosso

The Right to Food Guidelines, Democracy and Citizen Participation
Country case studies
Katharine S.E. Cresswell Riol

Peasants Negotiating a Global Policy Space
La Vía Campesina in the Committee on World Food Security
Ingeborg Gaarde

For further details please visit the series page on the Routledge website:
http://www.routledge.com/books/series/RSFSE/

Peasants Negotiating a Global Policy Space

La Vía Campesina in the Committee on World Food Security

Ingeborg Gaarde

LONDON AND NEW YORK

First published 2017
by Routledge

2 Park Square, Milton Park, Abingdon, Oxfordshire OX14 4RN
52 Vanderbilt Avenue, New York, NY 10017

Routledge is an imprint of the Taylor & Francis Group, an informa business

First issued in paperback 2019

British Library Cataloguing in Publication Data
A catalogue record for this book is available from the British Library

Library of Congress Cataloging in Publication Data
Names: Gaarde, Ingeborg, author.
Title: Peasants negotiating a global policy space : La Vía Campesina in the Committee on World Food Security / Ingeborg Gaarde.
Description: London ; New York : Routledge, 2017. | Series: Routledge studies in food, society and the environment | Includes bibliographical references and index.
Identifiers: LCCN 2016037772 | ISBN 9781138214873 (hbk) | ISBN 9781315444963 (ebk)
Subjects: LCSH: Vía Campesina (Organization) | Food and Agriculture Organization of the United Nations. Committee on World Food Security. | Peasants–Political activity–Developing countries. | Land reform–Developing countries. | Food sovereignty–Developing countries.
Classification: LCC HD1542 .G33 2017 | DDC 338.1/91724–dc23
LC record available at https://lccn.loc.gov/2016037772

ISBN: 978-1-138-21487-3 (hbk)
ISBN: 978-0-367-33509-0 (pbk)

Typeset in Bembo
by Taylor & Francis Books

Contents

List of illustrations viii
List of acronyms ix
Acknowledgements xi
Preface xiii
Foreword: From resistance to hope xv

Introduction 1

1 Exploring agency in a global movement 16

2 The consolidation of an international peasant movement 27

3 Carving out an autonomous space for participation in the UN Committee on World Food Security 46

4 The CFS as a political battlefield 65

5 From the plough to the UN negotiating table 95

6 Reaching outwards and looking inwards 115

7 Main tensions and debates: Strategic engagement with others 142

Final conclusions and perspectives: Peasants' agency in the global age society 168

Appendix A 181
Appendix B 183
Appendix C 184
Bibliography 194
Index 209

Illustrations

Figures

3.1 How the CSM engages with the CFS 52
A.1 Map of main field sites (2005–2015) 181

Table

2.1 Continuum of modes of organisation/operation 43

Boxes

3.1 Committee on World Food Security: key points of the 2009 reform document 50
4.1 CSM policies on investment in agriculture 70

List of acronyms

AIAB	Associazione Italiana per l'Agricoltura Biologica (La Vía Campesina member in Italy)
CFS	Committee for World Food Security
CFS-rai	CFS Principles on Responsible Agricultural Investments and food systems
CLOC	Coordinadora Latinoamericana de Organizaciones del Campo
CONSEA	Conselho Nacional de Segurança Alimentar e Nutricional (Brazil)
CSM	International Food Security and Nutrition Civil Society Mechanism
CSO	Civil society organization
FAO	Food and Agriculture Organization of the United Nations
FIAN	FoodFirst Information Action Network
HLPE	High Level Panel of Experts
GSF	Global Strategic Framework for Food Security and Nutrition
IAASTD	International Assessment of Agricultural Knowledge, Science and Technology for Development
ICC	International Coordinating Committee of the Vía Campesina
ICARRD	International Conference on Agrarian Reform and Rural Development
IFAD	International Fund for Agriculture Development
IFPRI	International Food Policy Research Institute
ILO	International Labour Organization
IMF	International Monetary Fund
IPC	International Planning Committee for Food Sovereignty
LVC	La Vía Campesina
LRAN	Land Research Action Network
MAELA	Movimiento Agroecológico de América Latina y el Caribe
MST	Movimento dos Trabalhadores Sem Terra (Vía Campesina member in Brazil)
NFU	National Farmers Union (Vía Campesina member in Canada)
OECD	Organization for Economic Cooperation and Development

ROPPA	Réseau des organizations paysannes et des producteurs agricoles d'Afrique de l'Ouest
SPI	Indonesia Peasant Union (Vía Campesina member in Indonesia)
TNI	Transnational Institute
UN	United Nations
UNAC	União Nacional de Camponeses (Vía Campesina member in Mozambique)
UNECA	United Nations Economic Commission for Africa
UNHRC	Human Rights Council of the United Nations
VGGT	Voluntary Guidelines on the Responsible Governance of Tenure of Land, Fisheries and Forest
WFS	World Food Summit (1996)
WTO	World Trade Organization

Acknowledgements

Throughout this course of research, I have benefited greatly from discussions and exchanges with social movement scholars from all around the world, as well as an uncountable number of activists engaged in a struggle for a fairer world. I express my gratitude to each and every one who inspired me and supported me along the way.

Special thanks go to Geoffrey Pleyers for being a source of inspiration and indispensable support during this writing process. While the list here is far from complete, I also wish to express my strongest gratitude for the dialogues I had with and support I received from Annette Desmarais, Josh Brem-Wilson, Nora McKeon, Philip McMichael, Priscilla Claeys, Jessica Duncan, Sergio Sauer, Breno Bringel, Flavia Braga Vieira, Matheuz Zanella and Katie Whiddon. Also, I would like to thank Jorge Chavez, University of Vale do Jequitinhonha, Minas Gerais and Marcos Piccin, the Rural Extension of University of Santa Maria, Rio Grande de Sul, for their warm welcome and help during my academic research in Brazil. Thanks to all professors and activists who accompanied me during field visits in West Africa, Brazil, and Indonesia. I also owe thanks to Hubert Ouédraogo (Lead Land Expert, Economic Commission for Africa (ECA)) who facilitated my research within the Network of West African Peasant and Agricultural Producers' Organizations (ROPPA) in West Africa and inspired me to continue my research on land tenure after my time in West Africa.

In regard to my research within the world of La Vía Campesina, I am deeply grateful to Nico Verhagen, technical support of the International Operational Secretariat of LVC, who supported me along the way, both through insightful discussions and from sharing of contacts with members of the movement around the world. This trust gave me a privileged position to understand and research a movement from within. The insights acquired of the values, strategies and organisational dynamics of La Vía Campesina largely come from dialogues with members of the movement conducted during research stays in different regions. In this respect, I am deeply grateful for the overwhelming openness I met while travelling alone in different corners of the world. During this research path, members of the movement, not only showed a strong willingness to meet and share with me, several peasant and farmers also invited me stay in their homes, from small-scale famers in West Africa, to youth activists in Indonesia, and settlement farmers in Brazil.

While every moment and encounter has been rich in different ways and has contributed to my understanding of the world of rural movements – particularly La Vía Campesina – I would like to give special thanks to Nettie Wiebe (La Vía Campesina, NFU Canada) Paul Nicholson (La Vía Campesina, COAG, Spain), Nettie Wiebe, NFU Canada, Renaldo Chingore João (La Vía Campesina UNAC Mozambique), Andrea Ferrante AIAB (La Vía Campesina Italy) and Maria Noel Salgado Spinatelli (MAELA, Uruguay) for their willingness to discuss and share with me.

In regard to my UN-based research, and in particular my time conducting research within the UN Committee on World Food Security, I am grateful to all members of the Civil Society Mechanism for receiving me so well in Rome. I also owe thanks to those government representatives and UN employees who accepted to participate in interviews and helped to open doors for me to do this research within UN arenas. Without the large number of people who were ready to share and support me along the way, this book could never have taken the depth that I insisted on.

Finally, I am running out of words to express my gratitude to my family and friends for all their love and support during the challenging process of writing the dissertation and subsequently to turn this manuscript into a book. The continuous encouragement from family members, new and old friends and dedicated academics around the world gave me the moral and intellectual support that has been essential to complete this book.

Preface

During the last ten years I have been straddling different worlds. From the field of peasants' struggles I have been working in solidarity with peasants and farmers seeking to transform and democratise the world's food systems. I have conducted research with and on peasants' networks in West Africa, peasant organizations in Indonesia, indigenous peoples in Mexico. I have engaged with landless workers and settlement farmers in Brazil and walked side by side with grassroots movements in many World Social Forums and people's summits around the world. From a decade of working in the network of civil society actors, as a freelancer and activist researcher, I gained insights into the struggles of people's movements occupying streets and fields in the fight for global justice. On the other side of the fence, I have been sitting side by side with policy-makers during my research in the UN headquarters in New York and the FAO headquarters in Rome. From within the corridors and official meetings in the UN buildings, I observed and engaged in dialogues with the global policy elite of government representatives' meetings in intergovernmental forums to decide on global strategies, principles and guidance for the future development path. The observations from the different contexts of grassroots everyday struggles and field research within selected UN arenas form the basis of this book.

Born on a small farm in the outskirts of Denmark, I was from a young age intrigued by how grassroots actors engaged in community struggles – and in particular farmers rooted in their territories – how they can act as key actors of global contestation? Curious to better understand the mobilization of the rural actors, during the combined food and energy crisis of 2009, I engaged in research with the coalition of farmers of 'the Network of West African Peasant and Agricultural Producers' Organizations' (ROPPA) to study their global mobilising around land grabbing. This research stay in Burkina Faso turned out to be a milestone in my journey with peasants' movements, in particular, due to my encounters with many members of La Vía Campesina: the global peasant movement bringing together more than 200 million farmers in 80 countries of the world. From my engagement with peasant networks in West Africa, I observed how multilingual teleconferences were set up to initiate active dialogue with other civil society organizations, State actors, local authorities and other partners. The network also engaged strongly in the preparations of the World Social Forum in Dakar 2011. With a strong impulse from the peasant movement, this Forum led to the civil actors organizations signing the "*Dakar Appeal Against Land Grabbing*". This appeal made a direct reference to the Committee on World Food Security (CFS) reformed in 2009 to include civil society actors in its policy processes. This link between peasants' struggles and this UN Commitee caught my attention. From my previous work with the peasants and indigenous people in Chiapas, Mexico, I learned about the social movements' autonomy and the difficult relations with state actors and institutions. Curious to explore how rural movements engage in intergovernmental negotiations, this led me to study how La Vía Campesina – without a much broader network – played a central role in crafting out a self-established autonomous mechanism for civil society organizations to engage in the CFS.

The book is anchored in a multi-site and multi-scale research design, combining the results of observations and dialogues of peasant delegates finding the way to international negotiations. It is based on fieldwork with different meetings with members of the movement in different settings on five continents. Based on the empirical findings, this book engages in a critical dialogue with the conventional wisdom about social movements' evolution and internationalization. Against the dominant view that social movements will either disappear or institutionalize in a predefined pattern, it examines the dynamics, dilemmas and debates arising from a global social movement engaging in global policy arenas.

Exploring how rural actors are negotiating their engagement in an intergovernmental world, this book seeks to shed light on what is behind 'the peasant voice'. It reveals the activists' own experiences, aspirations and personal thoughts on the challenging task of building transformative linkages between a formal UN arena and peasants'/peoples' struggles on the ground. With the strong focus on the capacity of peasants to act, this book is, however, not simply a tribute to La Vía Campesina, ignoring the internal contradictions and dilemmas within the movement. On the contrary, it seeks to shed light on the aspirations, challenges, dynamics and tensions that play out when a social movement takes the stuggle to the level of inter-governmental negotiations, without renouncing on its contentious stances and connections with its base. If this book contributes to a better understanding of these dynamics and challenges, and gives a glimpse into the existing opportunities for peasants and other citizens to shape the world from the local to the global, it will have achieved its aim.

Ingeborg Gaarde
(Rome, 25 June 2016)

Foreword: From resistance to hope

This fascinating book by Ingeborg Gaarde is at the intersection of two stories. One is a story of resistance; the other, a story of hope.

The first story has its departure point in the threats facing small-scale farmers, confronted with the advances of industrial forms of food production and the dominance of large agribusiness firms on increasingly globalized markets for food commodities.

Global peasantry has been destroyed, gradually but meticulously, by different means on different continents. In South Asia, the rapid spread of the Green Revolution since the mid-1960s bears a major responsibility. At the time is was launched, the Green Revolution saved lives: by combining the introduction of new, "high-yielding" varieties of plants, wheat and rice in particular, and by encouraging large-scale irrigation as well as the reliance on external inputs and on mechanization, it allowed yields to increase in massive proportions, in particular by making it possible to move to two or three harvests per year. But larger farmers, who could make the switch to a more capitalized form of agriculture, who had access to credit and could achieve economies of scale, benefited more than others. In contrast, many small-scale farmers working on the most marginal lands and relying on family labour suffered from the structural decline in farmgate prices combined with the increased pressure on land and water. To them, the Green Revolution meant that they could not remain profitably in farming: they abandoned their land and moved to cities; they offered to hire themselves as poorly paid farm laborers; many were relegated to subsistence agriculture.

Things were a bit different in Africa. There, the Green Revolution never took off, due both to the incompetence of governments – unable, or perhaps unwilling, to invest in proper rural infrastructures – and to development strategies that prioritized supplying cities with cheap food, even if that meant cheating farmers. It is Africa that Michael Lipton had in mind when, writing in 1977, he described the "urban bias" of the governing elites. And it was these elites' behavior that Robert H. Bates so vehemently denounced in the early 1980s, in books in which he described how the African farmers were the big losers of the overvaluing of local currencies, high export tariffs, and development projects that not only did not serve them, but had to be paid for at their expense, and at the cost of huge public debts. When, starting in the 1980s, structural adjustment programmes were imposed and began taking their toll on the ability of the African State to support farmers, international financiers could defend them, shamelessly but not entirely without justification, as the price to be paid for two decades of official mismanagement: the State had failed, let it fail, and let's focus on getting the "prices right", became the motto. The African farmer had been cheated, he was now left to fend for himself.

Latin America was a hybrid of both: it combined Asia's technological prowess (and it preceded Asia, in fact, in that regard) with the formation of alliances between big landowners and the bureaucrats – alliances directed against the *campesinos* and the landless peasants. The *latifundistas* and the *fazendeiros* of the continent benefited from an agrarian context in which the uneven distribution of land, in many cases hardly untouched since the colonial times, attained levels seen in no other region: they typically combined their economic dominance with political influence, to repress the peasant movement demanding land justice. (When I met the former president of Honduras Jose Manuel Zelaya Rosales in Managua in September 2009, just three months after he was deposed following a coup d'état, he attributed his removal from power to his intentions to launch a land redistribution – aware as he had become, he told me, of the impacts of rural inequality on the social fabric of the country. An heir of a big and powerful family of landowners himself, he was seen as a traitor by the members of the very party that had allowed his accession to power. In Latin America, it can be very costly indeed for a leader to turn into a reformist and to side with the struggle of peasants.)

It is against this background of large-scale neglect, overt repression, and gradual disappearance of the peasants in all regions, that the Vía Campesina was formed in 1993. More than a beginning, the establishment of the Vía was the outcome of a series of mobilizations on various continents – in Costa Rica, in Indonesia, in the Indian State of Karnataka. The message that was sent was a powerful one, particularly if we recall the background against which it emerged. The Uruguay Round of trade negotiations, conducted since 1986 under the framework of the General Agreement on Tariffs and Trade (GATT), was finally coming to a close: it would lead in 1994 to the establishment of the World Trade Organization and to the gradual submission of agriculture, through a new international agreement, to the disciplines of multilateral trade rules. Food, it was becoming clear, was set to become the next frontier of the great mill of commodification, and farmers from the world over were asked to compete against one another – and let the least competitive disappear.

The counter-offensive of the Vía Campesina could be summarized in the new expression they coined: they called their programme food sovereignty. Food sovereignty was, first and foremost, a story of solidarity against adversity, of cooperation against competition. The trade negotiators wanted their farmers to compete: instead, rallying behind the new slogan, the peasants of the world decided to unite. A strange ballet of words occurred: those talking about trade "liberalization" were condemning farmers to new forms of pressure and coercion from the global marketplace and from the large agrifood companies that dominate it, while those speaking of food "sovereignty" meant in fact the opposite of food wars – they meant alliances across national borders. With food sovereignty, a set of new displacements occurred: social movements replaced governments as the main source of legitimacy; the building of resilient communities through small-scale farming and the relocalization of agrifood systems was given priority over the search for efficiency gains and economies of scale; and (in the words of Jan Douwe van der Ploeg) the art of farming replaced the business of agriculture as the way to describe the future role of farmers.

Though increasingly popular within social movements, however, food sovereignty was only rarely endorsed by governments as a viable alternative to the combination of global trade and further industrialization of food production as a means to achieve food security. But then came the shock of the world food price crisis.

In the course of just a few months, between November 2007 and June 2008, the price of major food commodities on international markets suddenly sky-rocketed – the result of a combination of weather events and low stocks, high oil prices and massive speculation. The alternative programme of food sovereignty, until then derided by "serious" policymakers, suddenly had to be taken seriously. The crisis shed light on the impacts of the international division of labour that had developed since the 1980s, which saw low-income, agriculture-based countries caught in a trap. Because they had specialized in the production of export crops such as coffee, cocoa, tobacco or cotton, these countries had become increasingly dependent on imports to feed their growing populations, and their local farmers, although they were many, were unable to respond to respond to the price "signals" from the market: it was becoming clear that failing to invest in these farmers and in the production of food to meet local needs had been a tragic mistake, that had been going on for thirty years.

To reverse this, it had now become clear, two conditions had to be fulfilled. First, agricultural policies would need to achieve a much better balance between the support going to export-led agriculture and the support to small-scale food producers producing for themselves, their families and their communities. Supporting the local production of food crops now was defended not only as a way to reduce the dependency of the country on food imports, and thus the vulnerability of the country to price shocks on international markets; it also became to all obvious that it was the best means to raise incomes in rural areas, where, in low-income countries, the majority of the extremely poor still often reside. Second, jobs must be created in the industry and services sectors, in order to absorb the excess workforce migrating from rural areas. Ideally thus, what was required was a complementarity between these different sectors (agriculture, industry and services): small-scale, family agriculture should be supported in order to reduce rural poverty; but in addition, in what Irma Adelman famously called "agriculture-led industrialization", it may both ensure a market for the local producers of manufactured goods and service-providers, and should allow the growth of a food processing industry, and associated services, that contribute to the strengthening of local food systems.

It was to support countries in this radical redirection of their efforts in support of domestic food production that the Committee on World Food Security was revitalized. This is the second story on which Ingeborg Gaarde builds. It is the story of hope.

The reform of the Committee on World Food Security (CFS) is perhaps the single most significant development in the area of global food security in recent years. Initially established in 1974 following the first World Food Conference as an intergovernmental committee within the United Nations Food and Agriculture Organization (FAO), the CFS was reformed in 2009 with an aim to become "the foremost inclusive international and intergovernmental platform for a broad range of committed stakeholders to work together in a coordinated manner and in support of country-led processes towards the elimination of hunger and ensuring food security and nutrition for all human beings." This reformation was grounded in the recognition that governments will only manage to make true progress towards building sustainable food systems if they accept working in a bottom-up fashion, by learning not only from one another's experiences, but also from the experience of those who are on the frontline of combating hunger – the international agencies and the non-governmental organizations – and the victims of hunger themselves.

But, Ingeborg Gaarde asks us, shall the two stories harmoniously combine? Or shall they collide? If the Vía Campesina is to be co-opted into a process such as the CFS, does it mean that it shall lose its ability to invent new solutions? Shall it have to compromise? Shall "leaders" emerge within the organization, monopolizing the role of spokespersons, at the risk of making it more difficult for the members of the movement to be heard? According to a recent count, the Vía Campesina now brings together 164 local and national organizations across the world, scattered over more than 70 countries across Asia, Africa, Europe, and the Americas, and representing an estimated 200 million farmers: is it realistic to believe that the Vía shall escape the dangers of bureaucratization, if it is to be an effective participant in a UN process such as the CFS – one that requires expertise, the building of networks among the community of diplomats, an ability to articulate clearly and forcefully one's demands and, perhaps most challenging of all, an ability to strike compromises?

These are fundamental and difficult questions. Ingeborg Gaarde provides convincing answers, uniquely well informed both by her long companionship with those who shape the movement and by her participation in the processes that, at every turn, threaten to co-opt it – forcing it to lose its critical edge, its uncompromising denunciation of the dominant food systems, as the price to pay for remaining a credible actor on the CFS stage.

Beyond the Vía Campesina itself, the lessons she draws concern the movement towards food sovereignty as a whole. For food sovereignty, as we know, has now been embraced by a wide range of organizations across the world, far beyond small-scale farmers' organizations. Social movements have embraced it, as a means to redefine how food should be produced and commercialized. Environmental groups refer to it as a way to encourage agroecology and the relocalization of food systems – promising diets that are both healthier and more friendly to the planet. It is invoked by urban groups, who propose to establish food policy councils and to reconnect their city to its rural hinterland. The frontlines were the World Trade Organization ministerial summits in Seattle or Hong Kong; it is now the local school board, the company's canteen, or the local farmers' market. Alliances are now being built, at the local level, between citizens, farmers and municipalities.

Food sovereignty was accused of placing the interests of farmers above those of urban consumers: by some magic, it is now the urban middle class, often joining forces with low-income communities claiming for more food justice, that are the most dynamic part of the movement. Has is become mainstream? Certainly not. Has it proven its ability to be the source of promising alternatives, allowing us to dream of a world in which food will be produced differently, in which power relationships will be rebalanced, and in which even consumption patterns will be more sustainable? I believe it has. The decentralized invention of solutions, local experimentation with alternative food networks, not only serve to broaden the political imagination of policy-makers: they also allow the movement to participate in the search for solutions, moving away from a purely critical stance, without compromising its integrity.

I welcome the book of Ingeborg Gaarde as a major contribution to our understanding of the Vía Campesina, and more broadly, to the debate on the relationship between social movements and institutionalized processes of participation. The story of resistance, it turns out, ends with a message of hope.

Olivier De Schutter *is the former UN Special Rapporteur on the right to food (2008–2014). A member of the UN Committee on Economic, Social and Cultural Rights, he co-chairs the International Panel of Experts on Sustainable Food Systems (IPES-Food).*

Introduction

> Our greatest strengths are our self-reliance, our identity as peasant families, our diversity and our unity. These are taking us forward in our efforts to transform the food system.
>
> (Henry Saragih, Outgoing General Coordinator, Opening Ceremony at the Peasants' Victory Stadium, Jakarta, Indonesia, 13.06.2013)

This was the powerful message I encountered when I entered the opening ceremony of La Vía Campesina's 6th International Conference in Jakarta, Indonesia, June 2013. The old stadium hall in Jakarta was covered with banners of 'Globalise the hope, Globalise the struggle' and the conference hall was filled with emotions that ran through the stadium from the waving of flags of around 600 peasants and farmers gathered in Jakarta to celebrate two decades of building a global movement.

Ever since La Vía Campesina (LVC) was born in 1993 in Mons with the aspiration to become 'the international peasants' voice', LVC has been in rapid expansion. Today, LVC counts more than 200 million members, peasants, small and medium-size farmers, landless people, women farmers, indigenous people, migrants and agricultural workers from five regions of the world. The movement is increasingly recognised as what may be the largest global political movement on earth.[1]

Under the banner 'For land and people's sovereignty, in solidarity and struggle', LVC's 6th International Conference was organised around thematic panel discussions as well as a large number of parallel meetings and events. A separate group was created for what the movement calls their 'friends and allies'. In this meeting this group consisted of around twenty academics, selected NGO representatives, as well as members of global think thanks and journalists. The conference allowed all participants at the meeting supporting the movement from their different positions and corners of the world to intervene and present their suggestions in plenary sessions. However, voting remained strictly reserved to the members of La Vía Campesina (LVC, 2014a: 2). Whereas the gathering of representatives from different corners picking up a headset before being seated in a meeting hall may remind one of a UN meeting, the atmosphere of this peasant conference is no typical UN conference. In contrast to the formal character of most UN meetings, peasant delegates did not enter the meeting hall in a suit with a laptop case under their arms. Instead, delegates from around world were dressed in traditional colourful clothes, equipped with flags, music instruments, traditional drinks, food, seeds and other symbols of their struggle. Another feature of the peasant Assembly was that every morning session started with a *mística*: a symbolic event combining song, poetry, dance and the circulating of seeds. This vibrant gathering helped participants to better connect with each other and unite around their common struggle.

From morning to late evenings a group of interpreters supported the communication flow during the sessions from the small translator booths in the back of the room. These loyal and talented interpreters – most of them from the volunteer interpreters' network 'Babels'[2] – did not only provide translations into the five UN languages, but into no less than 15 languages. During the conference a special agro-ecological village was set up to display the diversity of farming methods and local techniques as an alternative to chemical and industrial agriculture. In order to present LVC delegates and allies with the local realities in the region where the meeting was held, a field visit was arranged to the Sinarjaya, an agro-ecological village established on a rubber plantation that had been occupied by landless peasants of the Serikat Petani Indonesia (SPI). During this fieldtrip, youth activists from SPI accompanied participants to the international conference in a long walk through their fields of crops. As guides during the march they proudly explained their engagement in the struggle for an agrarian reform promoting agroecology and a dignified livelihood for more than 1,000 peasant families (LVC, 2014a: 13). The day in the fields culminated with a symbolic ceremony of planting trees. All these elements gave the peasant conference a strong emotional and 'rooted' character.

Although the 6th International Conference in Jakarta was marked by the celebration of La Vía Campesina's 20 years of Peoples' action and solidarity (LVC, 2013a, b), this gathering was not only a commemoration. The Jakarta conference was a political meeting where women and men organised, discussed, shared, debated, and exchanged their views and experiences from their daily struggles. It concluded with the 'Jakarta Call', a collective declaration, conveying a clear message to rural and urban organisations of social movements to join together under the banner of 'Food sovereignty now – transforming our world':

> Today, we are facing a major crisis in our history, which is systemic. Food, labor, energy, economic, climate, ecological, ethical, social, political and institutional systems are collapsing in many parts of the world. The growing energy crisis in a context of fossil fuel depletion is being addressed with false solutions ranging from agrofuels to nuclear energy, which are among the greatest threats to life on earth. We reject capitalism, which is currently characterized by aggressive flows of financial and speculative capital into industrial agriculture, land and nature. This is generating huge land grabs and a brutal displacement of people from their land,

> destroying communities, cultures and ecosystems. It creates masses of economic migrants, climate refugees and unemployed, increasing existing inequalities.
>
> (LVC, 2013a).

The key messages from the conference were spread by the media team following the conference and key messages shared in writing and audio with all its members around the world.

The glimpses from the Jakarta Conference presented above reveal that LVC is a rural movement bringing together people who share a strong conviction: peasants – and other historically marginalised groups of society – can shape the world.

This book explores how LVC, since its foundation in 1993, has projected itself as a global actor and opened up political spaces[3] for its members to articulate their needs and transmit their knowledge, interests, and demands in their own voice (Desmarais, 2007: 23; Borras and Franco, 2009: 29).

The book focuses on how LVC has carved out a space for small-scale producers – and other citizens – to participate in UN global policy-making processes related to food. Before examining some of the opportunities and challenges to the engagement of and it members in the 2009 reformed Committee on World Food Security (CFS), the following sections present the main argument that will be developed throughout this book.

1 Global movements breaking out of the classical model of 'internationalisation with institutionalisation'

The common view among various analysts of social movements is that access to institutions leads to co-optation and de-radicalisation (Meyer, 1993; Piven and Cloward, 1977; Tarrow, 1998; Tilly, 2004; Kriesi, 1996; Kaldor, 2003).[4] Institutionalisation is largely portrayed as a negative drawback for social movements as it predicts that social movements will lose their spontaneity, moderate their claims, give up protest mobilisation, and in effect cease to exist as a movement (Tilly, 2004; Walker, 1994). Given the continued growth and impact of international networks in 1990, influential scholars foresee that internationalisation will be followed by the same linear model of institutionalisation (Tilly, 2004; Tarrow, 2005).

On an analytical level, one can distinguish between two – often simultaneous – processes: 1) *an internal dimension* whereby social movements are expected to develop from vibrant networks into rigid hierarchical and organisational structures with an increased gap between a professional elite and the member base; and 2) *an external dimension of institutionalisation* whereby the movement will decide to engage within an external entity, mainly to gain from coalitions with more resourceful actors, expecting the movement to become incorporated, 'tamed' and co-opted into the external structure.[5] In this book I ask: *Is this path avoidable?*

The main argument presented in this book is that if we wish to better understand today's comprehensive multi-scale activities of global movements, we need to move beyond dichotomies and towards less deterministic analytical frameworks than those that have largely dominated the literature of national social movements (Tarrow, 1998; Tilly, 2004). Social movements are not only responding to and are shaped by structures, but also contribute to defining and reshaping the arenas in which they engage at different scales (Godwin and Jasper, 2004: 2007–27; Pleyers 2012: 168; Giddens, 1984; Bringel, 2015; Jasper, 2010a).

These agency-structure dynamics will be analysed empirically when I explore LVC's engagement in the UN Committee on World Food Security. Through empirical research, I will demonstrate why we should be careful not to lock social actors too firmly into the policy arenas and processes in which they engage (Jasper, 2007). Rather than analysing the engagement of a social movement in a global policy arena from a simple 'success–failure' calculus, this book explores some of the tensions, debates, and reconfigurations arising from this engagement.

Finally, when I call for more a dynamic analytical framework to study global movements in this book, the aim is not to elaborate such a framework as such, but rather to show – through empirical fieldwork – that such patterns are not linear but rather complex, dynamic, debated, adjusted and negotiated.

The following sections briefly present the global context and processes around the 2009 CFS reform, which gathered speed when the global food crisis hit the media headlines in 2008–2009, leading to the awareness of a crisis in global food governance.

2 From global crisis to reform of the Committee of World Food Security

On 22 April 2008, the head of the United Nations World Food Programme (WFP) called for urgent action to tackle the 'Silent Tsunami' of rising food prices that threatened to push more than 100 million people worldwide into hunger (UN, 2008). The crisis in the global food system had reached the headlines with worrisome statistics of even more people being placed in the 'risk zone' of hunger due to the spikes in food prices (FAO, 2008). International media documenting and circulating the widespread waves of hunger protests (*The Guardian*, 2011) and revolts that consequently broke out in over 60 countries, affirmed that access to food was no longer only a moral imperative but also a security issue threatening national and regional stability (McKeon, 2015: 5).

Whereas public and political attention given to food security prior to the crisis had been practically 'negligible'[6] (Brem-Wilson, 2011: 203), the crisis urged governments to address issues of governance in the international food system (Duncan, 2015; McKeon, 2013). Among the number of initiatives that were launched as a response to the crisis was 'The Global Partnership for Agriculture,

Food Security and Nutrition' presented by French President Sarkozy under the Presidency of G8/G20[7] in June 2008. The declared aim of the Global Partnership was to 'bring together governments, regional bodies, international agencies, civil society, development banks and donors and businesses to develop coherent strategies against food insecurity'.[8]

A well-organised peasants' networks and broader society actors alliance mobilising around land issues, loudly criticised the idea of bringing food governance further into the hands of a global elite of leading governments, and criticised the proposal for lack of transparency and broad civil society consultation (GRAIN, 2011; Duncan, 2015; McKeon, 2013). While the civil society alliance lobbied against the proposal for the Global Partnership, member states in the Committee on World Food Security (CFS)[9] went through a drafting process leading to a different reform proposal to restructure the UN Committee into a multi-stakeholder institutional structure. In this reform proposal governments invited non-state actors to self-organise their full participation in the policy-process leading to global policy-making on food and nutrition, with an accent on addressing the needs of those most affected by food insecurity (McKeon, 2015: 109). This proposal to reform the CFS won over other proposals[10] and led the Committee through a wide-ranging reform process to make the CFS the 'foremost inclusive international and intergovernmental platform dealing with food security' (FAO, 2008). As McKeon, a Rome-based activist and author who was previously responsible for FAO–civil society relations put it in an eyewitness account of the 36th Annual Session of Committee on World Food Security:

> For the first time in the history of the UN system, representatives of small-scale food producers and other civil society organizations, along with private sector associations and other stakeholders, would be full participants and not just observers of the intergovernmental process.
>
> (McKeon, 2009b)

Through the CFS reform process, civil society organisations secured the right to autonomously co-ordinate their engagement in the Committee as official participants and are doing so through the International Food Security and Nutrition Civil Society Mechanism (CSM). One of the characteristics of this mechanism is its accent on those most affected by food insecurity, including small-scale farmers, fisherfolks and indigenous people – the groups of society that in the past have been largely dominated by international NGOs (CSM, 2013). Several activists interviewed for this book state that only a few years ago it would have been 'unthinkable' for rural movements to be active at this level. As stated by an LVC member during the 40th Annual session of the CFS:

> For a long time peasants did not know what was going on in international negotiations about food. These processes took place behind closed doors and decisions were restricted to governments, technocrats, bureaucrats and

the biggest food lobby industry. Now, in the CFS we speak on our own behalf and present our experience and contribute to designing food policies for the world.

(African-LVC member, CFS Annual Session, 40, Rome, 16.10.2013)

The next section shows how struggle over political leadership in the global governance arena is central to understand the emergence and entry of a rural movement into the transnational policy arena. In Chapter 2 will see how LVC under the slogan 'Nothing about us without us' has strived to become a more direct voice of peasants and small farmers than other groups tending to speak on their behalf: International NGOs and the International Federation of Agricultural Producers and NGOs (Borras and Franco, 2009: 30).

3 An international peasant movement seeking a voice

Since its foundation in 1993, LVC has strived to establish itself as 'the main intermediary between various local national movements of the landless peasants and small farmers at the global scale' (Borras and Franco, 2009: 30). As Basque farmer and one of the founders of La Vía Campesina, Paul Nicholson explained during the Second International Conference of the movement in 1996:

> To date, in all the global debates on agrarian policy, the peasant movement has been absent; we have not had a voice. The main reason for the very existence of the Vía Campesina is to be that voice and to speak out for the creation of a more just society...What is involved here is a threat to our regional identity and our traditions around food and our own regional economy...As those responsible for taking care of nature and life, we have a fundamental role to play...The Vía Campesina must defend the 'peasant way' of rural peoples.
>
> (Desmarais and Nicholson, 2013)

It is in this historical context that LVC has emerged to ensure that small and medium size farmers can speak in their voices about their own demands at the global level (Desmarais, 2007; Brem-Wilson, 2015). In the words of Desmarais:

> The goal of La Vía Campesina is to bring about change in the countryside – change that improves livelihoods, enhances local food production for local consumption, and opens up democratic spaces; change that empowers the people of the land with a great role, position, and stake in decision-making on issues that have an impact on their lives. The movement believes that this kind of change can occur only when local communities gain greater access to and control over local productive resources, and gain social and political power.[11]
>
> (Desmarais, 2007: 198)

The right to direct participation for peasants and small farmers aiming to have more impact on agricultural matters at all levels of policy-making has been a central component of the food sovereignty framework, launched by LVC in 1996. As stated under the principles of 'Democratic Control':

> Smallholder farmers must have direct input into formulating agricultural policies at all levels. The United Nations and related organizations will have to undergo a process of democratization to enable this to become a reality. Everyone has the right to honest, accurate information and open and democratic decision-making. These rights form the basis of good governance, accountability and equal participation in economic, political and social life, free from all forms of discrimination. Rural women, in particular, must be granted direct and active decision-making on food and rural issues.
>
> (La Vía Campesina, 11–17 November, 1996, Rome, Italy)

This pillar reveals that the food sovereignty framework requires a democratic, egalitarian approach to shape food policy at all levels (Patel, 2009: 670; Brem-Wilson, 2015: 1). It is with this goal of direct participation of food producer constituencies – and other citizens – at the transnational level that LVC seeks reform opportunities like the CFS reform to bring the direct voices and demands from rural people to the centre of global policy-making on food. In the words of a West African farmer's leader following the reform process from its early beginnings:

> Reforms at the level of international institutions like the CFS must be seized to advance the cause of peasants and family farmers. In truth, no policy whether at the national or international level truly takes into consideration the interests of this group, so it is up to us to say what we need.
>
> (Ibrahim Coulibaly, the National Confederation of Peasant Organisations/LVC, Rome 11.10. 2013)

Ultimately, La Vía Campesina's strong engagement in the CFS reform process must be seen as an aspiration to build *one* democratic single space for policy-making of food and agricultural global governance. Since its foundation, a main objective of LVC has been to remove agriculture from the World Trade Organization (WTO) – and the WTO from agriculture (LVC, 20006). The inspiration is that CFS can help in building a human rights based framework for governance in direct opposition to the WTO, which LVC perceives as an instrument of neoliberalisation (Borras, 2004; Desmarais, 2007). It is within this broader struggle over who has the right to design the global food system that this book explores how LVC is building its model of internationalisation, with a specific focus on the movement's engagement in the 2009 Reformed Committee on World Food Security.

a Peoples' organisations' linking to international institutions

The convergence of various crises – finance, food, energy and environment – put the need for coherent solutions to global hunger, climate and rural development on the global policy agenda. This has given way to a surge, maturation, and growth of global movements calling into question dominant development approaches and building alternatives by engaging in struggles over issues such as food, land, water, and livelihoods (Borras and Franco, 2009; Desmarais, 2007; McMichael, 2006, 2010, McKeon, 2015; Claeys, 2013). The surge of global peasant activism has led a number of scholars to observe how peasants and small-scale food producers' organisations are increasingly working at the global level while simultaneously building a model that relies on re-localisation and re-peasantisation (Desmarais, 2007; Borras, 2004; Borras and Franco, 2009; Edelman, 2003; Bringel, 2015; Rosset and Martínez, 2010; McMichael, 2006; Patel, 2009).

Activist scholars, especially, with strong connections to the movement, have shown how rural movements are increasingly getting involved in transnational networks, building spaces for peasant agriculture, and contributing to politicise the public sphere (Desmarais, 2007; Bringel, 2015; Borras and Franco, 2009: 29; Brem-Wilson, 2011; Patel, 2009). For instance, Borras and Franco (2009) argue that LVC has become a major protagonist in creating a 'citizenship space' for LVC as a distinct and direct voice of landless peasants. Another authoritative study of LVC is Annette Desmarais's account of LVC's first ten years (1993–2003). The author presents how the movement was built in opposition to neoliberal globalisation from the local to the regional and finally to the global level, in opposition to its main target, the World Trade Organization (Desmarais, 2007: 34). Building on existing literature and a developed multi-scale and multi-site research design, this book presents a concrete example of how mobilised peasants involved in struggles in a particular territory are increasingly capable of making local demands reaching the global governance level (Desmarais, 2007; McKeon, 2015).

This book contributes with analysis that discusses some of the opportunities, tensions and challenges for a social movement seeking to increase its leverage in the international arena while remaining a movement rooted in local and national struggles. Thanks to the methodological design combining participant observation from global meetings in the CFS arena with data collated within the movement in different corners of the world, this book goes beyond presenting the challenges for LVC as global protagonist. Exploring LVC as both an actor and an arena of debate among different national and sub-national peasant and farmers' groups (Borras and Franco, 2009: 11) the book goes beyond 'the peasant voice' by shedding light on actors' aspirations, demands and personal challenges. It shows how peasants activists today are building the infrastructure and facing the challenge of being 'translators' between grassroots struggles and sites of global governance.

b Global and local actions and dynamics

How to build the most efficient strategy for social change remains a main debate in the social movement literature (Pleyers, 2010; Piven and Cloward,

1997; Tilly, 2014; Tarrow, 1998). Rather than choosing one strategy, LVC seeks to build multiple strategies that allow for interaction, evolution and building of synergies between different forms and levels of political activism. Instead of repeating the slogan 'Think global – Act local' that many activists promoted in the 1980s LVC is an example of a movement seeking to build a strategy by being engaged simultaneously in global and local activism (McKeon, 2015: 113; Pleyers, 2010: 202; see also Desmarais, 2007). The movement is building a path where global engagement is not opposed to the local, but rather seeks to build synergies between different scales of activism to respond to global challenges.

While LVC continues to be an expanding network engaged in direct action and mobilisations outside the halls and against various international elite fora and global institutions such as the G8 summits, the World Bank and the World Trade Organization, the movement is increasingly engaging in policy institutions that address issues related to food, nutrition and agriculture. The increased engagement of rural activists in global policy dialogues may be a sign of a heightened awareness that peoples' struggles built from the 'bottom-up' may also be supported from the 'top-down' (McKeon, 2015: 85; see also Brem-Wilson, 2015). La Vía Campesina expressed in a document evaluating the first two years of engagement in the CFS that 'It is no longer simply a question of action inside or outside, but rather of thinking of the most efficient strategy to combine all the possible levels of action in order to reach our goals' (LVC, 2012a: 14). In the context of the CFS, civil society's demand from the very outset has been that the reformed CFS should aim to be a new platform to support struggles 'on the ground':

> A renewed CFS has to have strong links to the local, national and regional level, in order to contribute to assisting the struggles of actors to open up spaces of governance and policy-making at those levels, and to ensure that the perspectives and knowledge of locally based actors is heard and acted upon.
> (CFS, 2010).[12]

As Chapter 2 of this book will discuss more in detail, LVC, as a global movement, was born from an understanding that peasants in isolation cannot solve their common problems. As stated by one Asian LVC member at the Jakarta Conference:

> We must avoid the risk of being isolated. Instead, we must find a way to globalise the struggle to confront the forces behind the neoliberal model everywhere. Grassroots struggles are important but often ignored. We need to occupy spaces to confront power and build our struggles at all levels.
> (LVC member LVC 6th International Conference, Jakarta 10.06.2013)

Whilst the global (policy) arena has become an increasingly significant venue of action for LVC, it does not automatically mean a shift away from a local, national or regional focus.[13] In the words of an African peasant farmer, engaged in global policy work in the CFS:

> When we say, that we must always think globally, it doesn't mean that we forget that the struggle to be taken is at a local and national level. We need to build the peasant struggle to pressure everywhere and gain more on our side.
>
> (African LVC member, interview, Rome, 05.10.14)

This statement resonates with the outlook of many LVC activists, namely to 'bring the local in the global and the global in the local'.[14] This vision was also presented as a key objective in the report coming out of LVC's 6th International Conference in Jakarta:

> We maintain the visibility of local struggles globally and at the same time highlight the global aspects in local struggles. This is even more important in our efforts to build alternatives, as we are building ecological alternatives and food sovereignty that start from local action and are solutions to global problems.
>
> (LVC, 2014a)

This vision shows how the aspects of activities of social movements cannot be studied in isolation from the local and vice versa; the local or national dimensions increasingly require us to apply a global outlook (Desmarais, 2007; Beck and Cronin, 2006; Bringel and Domingues, 2015; Pleyers, 2015; Wieviorka, 2014; Wieviorka and Calhoun, 2013). As Wieviorka and Calhoun express it: 'understanding today's actors and challenges, we need to combine scales of action and levels of analysis, from the local to the global, and from personal subjectivity to globalization' (Wieviorka and Calhoun, 2013).

c A social change strategy building confluence between two paths

This book seeks to uncover some of the opportunities, challenges, and limits related to LVC's seeking to found its vision for social change by building convergences and synergies between different forms and levels of policy activism: from the everyday peasants' struggles to the level of international negotiations in selected global arenas.

Exploring LVC's engagement in an intergovernmental UN arena demonstrates how the movement is both engaging *horizontally* in cross-border alliances with other social movements and civil society networks, while simultaneously building greater *vertical* integration (Borras and Franco, 2009: 9). Seeking to build a strategy that links different levels of actions, La Vía Campesina aligns its vision for social change with the strategy presented by Edwards and Gaventa:

> Global citizen action implies action at multiple levels – local, national and international – which must be linked through effective vertical alliances.

> The most effective and sustainable forms of global citizen action are linked to constituency building and action at the local, national and regional levels. It is equally important that such action be vertically aligned so that each level re-enforces the other.
>
> (Edwards and Gaventa, 2001: 281)

This book suggests that LVC's strategy to combine different forms of political activism from the peasants' day-to-day struggles with alliance building advocacy work in international institutions can be seen as a confluence of the two logics identified by social movement scholar, Geoffrey Pleyers (2012). In his work *Alter-globalization* (2010), Pleyers differentiates between two prototypes of activist seeking different strategies to oppose neoliberal globalisation; what he calls the 'the way of reason' and 'the way of subjectivity'. Activists belonging to the first category tend to ally with international experts in international organisations and lobby in negotiations to push for global social change. Their approach is mainly 'top-down'. These types of activists, Pleyers argues, tend to challenge neoliberalism by seeking to reshape and monitor international institutions, by lobbying policy-makers and by drawing up expert reports. They use their energy mostly at the global policy level, where they seek to build norms and discourses and influence multilateral policy processes. They often see the locally rooted dynamics merely as a stepping-stone on the way to the main challenge, which is situated at the global level (Pleyers, 2010; Chapter 3).

The other group of activists, Pleyers argues, often rejects engaging with states and institutional actors, as these are seen as illegitimate interlocutors. Social movements acting from a logic of subjectivity derive their energy from grassroots activities, such as community building, in places where they can build more concrete alternatives and enact more visible social change. Activists adhering to this logic, the author argues, often ignore the importance that the global level may have on local conditions whereas the larger-scale transformation remains a blind spot (Pleyers, 2010; Chapter 2).

LVC's seeking to build its model of internationalisation by increasingly making use of global policy arenas to build support for the movements' struggles also resonates with Keck and Sikkink's 'boomerang model'. This theory outlines how place-based movements and transnational networks can use transnational activism to pressure national governments to change their practices to align with international norms (Keck and Sikkink, 1998: 13). We return to how the opportunities and challenges related to the building of such productive global–local linkages play out in practice in Chapter 4 and Chapter 6.

The following section presents the structure that will guide us through this book.

4 The structure of the book

Chapter 1 presents the approach to social movements adopted in this book and how the book seeks to bring social actors to the centre of the analysis by exploring the concept of *agency*.[15] The chapter motivates the choice of a multi-scale and multi-site research design to study LVC, as both an actor and an arena (Borras, 2004: 3; Borras and Franco, 2009: 1). It reveals how this research evolved while presenting some of the challenges related to conducting field research in a constant 'back and forth' between UN arenas and at global and local meetings of farmers' movements. It also present a number of reflections on the researcher–movement relationships from closely studying a social movement, in particular the challenges of getting access to research close to peasant activists and the importance of trust building. The chapter concludes with a discussion on the potential that lies behind the relationships between social movements and scholars, including the challenges faced in building these relationships.

Chapter 2 uses existing literature to explain how La Vía Campesina came together in 1993, as well as the movement's organising principles and values. These important components, along with the distinct politicised peasant identity, help us to understand the movement's political culture and what it is that brings members to unite in diversity. It illustrates how the movement has built itself in strong opposition to neoliberal top-down autocratic structures and is consequently organised as a 'people's organisation' (PO). The political culture of LVC entails characteristics and values that place the movement at a low degree of institutionalisation (e.g. strong focus on inclusion, participation, autonomy, horizontality, process). The chapter concludes that the organisational model depends to a high degree on trust. It is an attempt to build a middle way between 'loose networks' and top-down structures and bureaucratic hierarchies, which have haunted other social movements.

Chapter 3 presents how LVC has played a key role in carving out an autonomous space for civil society participation in the Committee on World Food Security (CFS). More specifically, the chapter presents the establishment of the International Food Security and Nutrition Civil Society Mechanism (CSM) as an interface with the CFS (LVC, 2012a: 13). This institutional innovation is an example of civil society actors finding ways to combine claims of autonomy with international engagement as well as seeking to bring those most affected by food insecurity to the leadership of policy processes. While these remain ongoing struggles, the innovations around the CSM enables resistance of the classical model of institutionalisation, affirming that processes and spaces are not pre-fixed or given, but rather shaped to a high degree by the actors engaging within them.

Chapter 4 explores the CFS as a political arena where deliberation, negotiation and confrontation take place. The chapter explores some of the unfolding dynamics arising from rural social movements' engagement in the CFS through the CSM. In order to reveal some of the dynamics arising from social movements in the CFS policy arena, the chapter mainly presents examples from participant observation conducted during the preparations and negotiations over the CFS Principles on Responsible Agricultural Investments and Food Systems (CFS-rai) that culminated with two rounds of negotiations in May and August 2014.

The chapter also demonstrates how food sovereignty as a political framework helps members of the CSM to negotiate in the CFS. It shows how this framework helps to guide the political action and to shape a coherent vision for the future of food security that builds on the right of peoples, communities, and countries to define their own food system and agricultural policies. The chapter argues that, beyond the struggle to win support for policy proposals that can have a positive effect on social change on the ground, the engagement of radial social movements in the CFS is also a struggle over identity, knowledge and meaning-making, as well as a battle over who has the right to be included in deciding these meanings (Alvarez et al., 1998).[16] Such observations call for a renewed and more nuanced debate about what successful engagement of social movements' participation in a multi-lateral system means. The chapter ends with considerations of how strategic engagement plays out from within the CSM space, and how direct participation of those most affected by food insecurity in global policy-making on food brings new issues to the debate about legitimacy, representativity and democratising of global policy-making.

Chapter 5 focuses on how the representatives of rural constituencies experience their own participation in the CFS. The chapter explores some of the potentials and constraints for agency arising from the actors' attempts to confront and alter the structures by engaging within them. It sheds light on interaction between those social movement activists seeking to become an actor (through participation and voice) within such processes, and the conditions that have to be attained for this to happen. In particular, the chapter reveals some of the challenges for members of rural constituencies – and other social movements – to translate formal rights into 'full' participation in a formal UN Committee. A main challenge that has been identified regarding social movements' engagement in the UN Committee is how to accommodate the dominant language, codes, speed and rhythms as well as the output-driven 'efficiency logic' of the UN. The chapter ends with a call for more research on the structure–agency dynamics and patterns of inclusion/exclusion of social movements' participation in global policy processes.

Chapter 6 shows how peasants who are delegated to the CFS strive to resist some of the institutional dynamics that have been predicted by one the most dominant models of social movements, i.e. institutionalisation. The common view among many analysts of social movements is that access to institutions leads to co-optation and de-radicalisation as activists are expected to modify their claims to ones that are more acceptable to authorities (Meyer, 1993; Piven and Cloward, 1977; Tarrow, 1998; Kriesi, 1996; Kaldor, 2003). In the dominant literature, institutionalisation is thus largely presented as a negative drawback for social movements, as it predicts that they will lose their autonomy and spontaneity, only to moderate their claims and ultimately give up protest and mobilisation. The dominant theories around social movements also predict a process of professionalisation, where leaders will uncritically be drawn into the international system, leading to a 'grassroots elite' increasingly getting cut off from their base. By uncovering the attitudes and practices of the peasant delegates that participate in the CFS, the chapter shows that another model of engagement is possible. Besides using their time seeking to develop the local–global infrastructure between grassroots struggles and sites of global governance, rural members of LVC engaged in the CFS also use considerable time and energy to look inwards: to conduct self-assessments and address issues such as accountability and representativity. The chapter ends with a discussion of some of the limitations of implementing this global–local strategy in practice.

Chapter 7 explores LVC's strategy as a social movement seeking to engage with others while jealously protecting its autonomy. In order to demonstrate this balancing act, the chapter presents examples of LVC's (at times uneasy) relationship with 1) international institutions, 2) States, and 3) NGOs. The chapter presents different visions of social change and shows how LVC is an arena of exchange between diverse national, sub-national peasant and farmer groups, who are experimenting with building their own strategies (Borras, 2004: 3). The chapter suggests that LVC remains a movement with room for tensions – expressed primarily though conflict and debate among activists themselves – which may help the movement to adapt to new situations.

The last section of the book presents how mobilised peasants seeking to build the local–global infrastructure between peoples'/peasants' struggles and engagement in global policy arenas have become key actors and producers of the global age society (Albrow, 1996; Pleyers, 2010). The comprehensive multi-scale activities of global rural movements compel us to move towards a more dynamic analytical framework to understand today's social movements. In particular, the evidence of peasants/peoples' organisations acting across multiple scales and arenas may be an indicator of an upsurge of new kinds of agrarian activism (Edelman, 2011: 85; Borras and Franco, 2009; Desmarais, 2007).

The closing section briefly discusses how the CFS arena today, with the direct participation of representatives of peasants/peoples' organisations in UN negotiations, is an example of a more complex, inclusive and democratic foundation for global policy-making. The book closes with a discussion of possible further research perspectives to examine the future direction of global food governance. The transformative potential that lies in building synergies between the worlds of social movements, academia and international institutions will have a central place in this discussion.

Notes

1 *The Guardian*, 17 June 2013: http://www.theguardian.com/global-development/2013/jun/17/la-via-campesina-henry-saragih
2 Babel is network of volunteer interpreters that was originally put together to 'cover the interpreting needs of the Social Forums' and of other international events, as defined in the calendar of the call of the social movements: http://www.babels.org/spip.php?rubrique2
3 With 'space' in this book I refer to processes or venues through or within which peasants engage in shaping policy processes related to food.

4 Other examples of scholars predicting a linear pattern of social movements' institutionalisation are: Stammers, 2009; Touraine, 1973; Walker, 1994; Gamson, 1990; Della Porta and Diani, 1999.
5 These two dimensions of institutionalisation are adapted from Pleyers (2012) demonstrating how the World Social Forums evolved towards of pattern of 'internationalisation without institutionalisation'.
6 While numerous international summits, particularly in 1974, 1996, and 2002, addressed the question of global hunger periodic commitments (McKeon 2009a), during the years approaching what in the official discourse is often referred to as the "2007–2008 food price crisis", the amount of political attention to the issue of hunger was low and declining (FAO, 2008: 4).
7 G8, or 'Group of 8', is a forum for the eight most industrialised countries in the world.
8 un-foodsecurity.org/node/135
9 In particular, Brazil and Argentina, who were supported by a number of African and EU countries. According to Brem-Wilson's account of the CFS reform, the Argentinean CFS Chair, Maria Del Carmen Squeff, apparently played a key role in opening up the policy process to non-governmental actors (Brem-Wilson, 2011: 210).
10 In his detailed analysis of the CFS reform, Brem-Wilson argues that this set of unique factors led to favourable conditions for reforming the CFS. One factor was the invitation by the CFS Chair for civil society to participate extensively in the reform process. Another factor was that the Food and Agriculture Organization of the United Nations (FAO) was itself emerging from an extensive process to reform the institution, in the context of the food price crisis. These contextual factors contributed to the member state proposal to reform the CFS, giving rise to a very dynamic and ambitious outcome (Brem-Wilson, 2015: 15).
11 This goal was later expressed in the official LVC policy document coming out of La Vía Campesina's 5th International Conference in Maputo, Mozambique in 2008 (LVC, 2008: 141).
12 Outcome document of the 2009 People's Food Sovereignty Forum in Rome "A discussion about civil society participation within the World Committee of Food Security" CFS 2010/9 Annex 3. (The key principles coming out of the working group on civil society participation within the CFS; background document for civil society organisations, not endorsed by the CFS.)
13 An example of peasants building complex patterns of collective action is the foundation of the Coordinadora Latinoamericana de Organizaciones del Campo (CLOC) in 1994. From this regional base in Latin American, the CLOC is an example of rural movements seeking to gain political influence and legitimate their claims by combining social action at the local, national, regional and international scale (Perreault cited in Rosset and Martínez, 2010: 156; see also Bringel, 2015 and Fox, 1994).
14 This was, for instance, articulated by an Indonesian youth member of LVC, Jakarta, interview, 13.06.2013.
15 Agency is broadly understood in this book as the will and capacity to act and intervene in the world (Giddens, 1984; Pleyers, 2015)
16 Alvarez et al., 1988 examined similar issues, in particular in the context of social movements in Latin America, where the authors refer to the cultural politics of social movements as a key concept for contestation and struggle over meaning-making.

1 Exploring agency in a global movement

This chapter presents the approach and the design applied to study LVC in this book. In the first part, I present how I apply an approach surrounding 'agency', with the aim of analysing La Vía Campesina as both an actor and an arena (Borras and Franco, 2009: 11). In the second part, I present how the research evolved and the challenges I faced in order to conduct a multi-scale and multi-arena research design, based on 130 interviews conducted in five different languages in various countries. The section ends with a number of methodological considerations, reflections on the data collection, the research design, as well as challenges related to the researcher–scholar relationship.

1 Studying a global movement

a La Vía Campesina as actor and an arena

La Vía Campesina is constituted by diverse autonomous member organisations such as peasants, small and medium-size farmers, landless people, women farmers, indigenous people, migrants and agricultural workers, who organise events, campaigns, and exchanges around the world. It means that the movement is not only a global *actor*, but also an *arena* consisting of various subspaces where debates and exchanges take place between heterogeneous actors from different social classes, ethnicities and political viewpoints (Borras, 2004: 3; Borras and Franco, 2009: 11).[1]

Wieviorka argues that the 'unity' of the movement should not be confused with the existence of a single organisation encompassing its various components (Wieviorka, 2005).[2] The unity of the movement relies on the coherence of social meanings shared by the actors who embody them (Touraine, 1978; Melucci, 1996). In a similar vein of thought, Pleyers maintains, the movement is constituted by the coherence of meaning that is embodied and shared by the social actors, 'the unity of the movement is not in the least incompatible with the heterogeneity of its actors' (Pleyers, 2010: 11). It is at this analytical level that the unity of the movement can be grasped, when we talk about LVC as a global heterogeneous movement.

It is from this understanding of the movement, that in this book I explore LVC as both an actor and an arena. Studying the movement, as an arena or as a 'multi-layered actor' (Pleyers, 2010: 207) does not only mean going beyond LVC's engagement in international meetings, summits, specific events and initiatives where the movement acts as a global actor. It also means digging into the cultural aspects of the movement,[3] the movement's social change vision, the personal dimensions of activism (Jasper, 2010b; Wieviorka, 2012; Pleyers, 2010; McDonald, 2006) and lastly, to shed some light on the movement's underlying contradictions and creative tensions (Touraine, 1981; Melucci, 1996).

b Social movements as a heuristic tool: an emphasis on agency

The French movement scholar, Alain Touraine, states that a social movement is defined by three constituent elements: the actor, a clearly defined opponent and a social conflict (Touraine, 1985: 772).[4] Touraine, and other social movement scholars with focus on the subject have stated how social movements provide us with a 'heuristic tool', i.e. that social movements help to understand both the current society and elements of the emerging society (Melucci, 1996: 25; Touraine 1973; Wieviorka, 2012; Pleyers, 2010).

The authors behind this tradition of social movement studies with social actors at the heart of their analysis often oppose themselves to the sociological paradigms of what Touraine calls 'classical sociology' or 'sociology of order' referring mainly to functionalism but also to the structural determinism of Marxism (Touraine, 1981: 34–35). Touraine states that the aim of a social movement is not simply to react to existing structures (e.g. existing inequalities), but rather to try to change the norms and values of cultural and social life (Touraine, 1995: 240). According to Touraine the dominant sociological paradigms leave little room for social action and thus for the possibility of transformation of society itself.

In order to get a better understanding of LVC and what drives the social actors constituting the movement, I apply the concept 'agency'. Giddens defines agency as 'the capacity to act otherwise…to intervene in the world or to refrain from such intervention, with the effect of influencing a specific process or state of affairs' (1984: 14). In addition to integrating Giddens' definition, the conceptualisation of agency in this book is inspired by Pleyers' approach to agency: 'The will to become an actor…and to have an impact on the way our common global future is shaped' (2010: 11).[5] My understanding of agency therefore contrasts with the commonly used 'agent' in political science theory, economics and game theory, which signifies a person who is acting on someone else's behalf.[6]

This book's focus on agency does not suggest that agency is 'free-floating'. Rather, agency consists of structures that can be simultaneously transformed whilst acting within them. According to Giddens' theory of structuration, human agency and social structure are interlinked. The author contends that processes and spaces are not pre-fixed. Giddens states that it is the repetition of the acts of individual agents, which reproduces the structure (Giddens, 1984; Giddens and Pierson, 1998: 77). This means recognising that there is a social structure, which is composed of traditions, institutions, moral codes, and established ways of doing things. However, it also means that these can be changed when people start to ignore them, replace them, or reproduce them differently (Gaunlett, 2002; Giddens, 1984).

The notion of agency applied here is also inspired by Touraine's focus on social movements: the will of social actors to shape not only their own life situation, but to shape society in a broader political struggle over values and gain control over what Touraine calls 'historicity of society' (Touraine, 1973: 354).

Bringing the actors into the heart of the analysis has helped me to delve deeper and to develop an understanding of how social actors build their organisations, networks, strategies as well as the ongoing tensions and debates that drive the movement (Pleyers, 2010: 13). It has helped me to better capture the 'system of meanings', in Touraine's words, i.e. the projects, values and visions that guide the actors and the orientations of the movement (Touraine, 1981: 271).

In this book, I endeavour to explore social agency by shifting the actors to the forefront of my analysis. This helps us to examine questions such as: Who are these social actors seeking to transform the global food system and to advance their struggle in the United Nations (UN)? What drives rooted peasants to travel to global meetings and negotiations to challenge global policy-makers? How do the social actors experience and make sense of conducting these shifts from the plough to the negotiation table at the UN? What challenges do they face and how do they seek to meet them? Can we learn something about society by looking at these actors?

The following section presents some of the methodological challenges to apply a multi-site and multi-layered research design for this book (Pleyers, 2010).

2 A multi-site and multi-layered research design

In order to explore LVC and its engagement in policy processes in a UN Committee from an academic perspective, I identified the main literature to shed light on LVC and gain new insights in social movement literature. Due to my work experience and my multidisciplinary background (trained as a political scientist with specialisation in the United Nations and on-the-ground experience with networks of rural movements and international NGOs in the UN milieu), I was already familiar with the civil society literature and the debates related to the United Nations. Combining fieldwork and theoretical explorations throughout my work meant a constant back and forth between the field and 'desk work', analysing interviews, documents, reports, statements, hand-outs and field notes from participation in LVC meetings, UN negotiations, UN side-events, etc. This alternating between the field and construction of an analysis helped to build a better understanding of the reality of the movement (Pleyers, 2010: 106).

In contrast to more theoretical methods in which the collection of empirical data is guided by preliminary theoretical exploration of what to expect, this approach has been overall inductive (Kaufmann, 1996: 22) and has relied a great part on empirical fieldwork. This means that rather than spending my first year focusing on my literature review, and then spending another year doing fieldwork, I combined fieldwork and theoretical explorations throughout the course of the research.

This approach also gave a flexibility to seize fieldwork opportunities as they emerged and to accept invitations to present and discuss my research in different contexts. The data collection and analysis greatly benefited from my participation in a number of international academic conferences. In 2012, when I was invited by the University Vale de Jequitinhonha (North-Eastern part of Minas Gerais State of Brazil) to present my work, I combined it with fieldwork in the region. In this regard, university professors and other activists in the LVC milieu have been very helpful in facilitating and accompanying me to the field in visits in the outskirts of cities in Brazil. This led to field visits and stays with small farmers, Indigenous peoples and Quilombola communities, as well as with MST[7] families in settlement camps. During these exchanges I got the chance to speak with members of LVC about my experience following the movement's engagement in the World Committee of Food Security. This way, I expanded my empirical data by discussing with other LVC members that are not directly engaged in the global policy work.

Overall, this approach enabled me to deepen my understanding of the realities of the heterogeneous member base constituting LVC as a global movement. In Appendix 1 I present a detailed account of seminars and conferences I participated in during the course of my research. The following section presents reflections on how this research design included a number of challenges.

a Getting access and the importance of trust

Those wishing to examine the functioning and the strategies of social movements are often faced with the critical challenge that data collection – and the quality of the analysis – often depends on getting access to the 'object' of a study (Edelman, 2009). During this research I experienced how seeking access to LVC's formal and informal meetings – where debates and tensions are often played out – implied a strong methodological reliance on contacts to movement leaders (Borras, 2008b: 30).

When this research took speed in 2012, I contacted a number of leaders from LVC, whom I knew from my background of working within the network of farmers organisations in West Africa. This led me to conduct an early Skype meeting with a technical staff member of LVC. In this conversation we discussed my research and my possible participation in LVC's 6th International Conference, the movement's highest decision-making body, to be held in Jakarta, in June 2013. The LVC staff member informed me that the decision of whether I could attend the International Conference had to be made by the International Coordination Committee (the ICC). I was therefore asked to write a detailed letter to the ICC in three languages, including my CV, my research agenda, the list of people that I had already interviewed, a full list of everyone I had previously worked for and the names of previous professors and employers. My request to conduct research inside LVC subsequently went through a consultative process within the movement and was finally accepted by the International Coordination Committee of LVC.

The invitation to participate in LVC's 6th International Conference in Jakarta the following year (June 2013) in many ways became a turning point in this research. A few weeks before the event, I received the official invitation to participate in LVC's 6th International Conference. This invitation letter spelled out the different roles of 'observers' and 'participants', the latter reserved to members of the movement. The group of observers included academics, selected NGOs, media persons, as well as a few invited speakers from certain UN agencies.[8] When I arrived in Jakarta, I felt deeply honoured to find myself in the 'Friends and Allies Group' consisting of a number of world known scholars[9] and leaders of international NGOs supportive of the movement's goal.

During the Jakarta conference, I was met with a remarkable level of openness from all participants. LVC members from different corners of the world invited me to visit their organisations, participate in their meetings and to stay with them in their regions. I experienced that having history of already working with members of La Vía Campesina from my research in West Africa, mainly from my position as a researcher within the regional Network of West African Peasant and Agricultural Producers' Organizations (ROPPA) in West Africa, played a supportive role in gaining access and establishing trust within the movement.

After being granted access to participate in the Jakarta conference, I was welcomed to do my research within the movement without any constraints on my research agenda. I was aware of the movements' concern for its autonomy (and reluctance to engage with young researchers) and expected more restrictions. Instead, I experienced how building trust with members in the movement meant that I was granted information and access to documents and meetings as an 'entrusted researcher'. Simply reviewing publications and online materials would certainly not have provided me with the valuable insight into the debates, tensions and dynamics within the movement.

b Shifting field sites

Conducting fieldwork in various different venues where LVC members were engaged required an extensive effort to understand how these different sites work. At times, painful efforts were made in order to make academic sense of these observations.

For instance, studying LVC's engagement in UN policy meetings and negotiations meant not only that I had to break down the complex processes, code the UN language, and analyse official statements and texts. It also involved attempts to uncover interworking, power relations and the different logics that guide the social actors' actions in this policy arena. As Müller observes, seeking to accomplish this type of participant observation, the researcher is faced with the challenge to obtain 'intimate knowledge' of the organisations and actors in a field that is 'ephemeral, dispread, complex and often opaque' (2013: 4). While these interactions and processes can thus be very subtle and difficult to graph, these must be uncovered by the researcher's capacity to mobilise his/her skills to carefully observe interactions, including 'those that seemed mundane and ordinary as well as others heavily loaded with symbolic meanings and issues of power' (Müller, 2013: 4).

Besides the need for attentive observations in the dense power field of the UN intergovernmental world, I experienced how conducting research in a frenetic UN work-environment requires a good deal of patience and flexibility. Since my first observations of UN policy processes at the UN headquarters in 2008, I had experienced how conducting research within a diplomat-dominated world means that you must build your own methods and techniques to obtain the information you need for your research. For instance, when interviewing UN diplomats you are often faced with closed doors and auto-responses such as 'I need to refer back to Capital'. Interviewing is therefore often improvised.

Researching within the world of the UN taught me the advantage of conducting informal interviews (e.g. not putting the voice recorder on the table) and the purpose has been to get more honest and direct answers, for instance about negotiations and how UN diplomacy works behind the scenes. Consequently, many interviews and participant observations for this book have been conducted in very different settings, from UN intergovernmental negotiations, corridor meetings, informal meetings during lunch and dinner, to bus rides and other types of personal encounters. Conducting research with peasants in the local environments and in the UN arena populated with UN diplomats meant social skills and considerable effort to adjust to different cultural contexts, rules, norms and language codes when moving from one research field arena to another.

In particular, I experienced how conducting this type of anthropological research close to social movements' activists meant constant reflection on the variety of roles and statuses held by the researcher in different contexts. Particular sensitivity was paid to differences in age, gender, class, culture and religion, as these may raise ethical issues.[10] For instance, in a discussion with an elderly food activist after a CFS session in Rome, I realised that in applying the same questions and interview style that I had developed for interviews with UN representatives and UN officials, I unintentionally offended this elderly food activist.[11]

This experience early on in the research process was an important lesson in the constant learning curve to find my own way to manoeuvre between different field sites and constantly reflect on my own role as a researcher.

c Researcher–activist relationships and ethics

During the course of my research, I experienced how conducting research close to social movement activists is related to a set of ethical concerns. First of all, being an 'entrusted' researcher participating in formal and informal meetings within a social movement and its engagement in a UN arena means collecting data that must subsequently be treated carefully. In particular, conducting this research within LVC, which is known for being a movement that zealously cares for its autonomy, forced me to reflect on the delicate role of being an 'entrusted researcher'.

Conscious of the fact that I was entrusted not to share strategic and sensitive information,[12] I felt in some situations that the openness and amount of information shared with me by LVC activists was almost overwhelming. On several occasions during this research, I found myself confronted with the uncomfortable feeling of studying 'sideways', i.e. writing about personal meetings and the observations of people I was surrounded by. I tackled this challenge by looking into the social movements–scholar activist literature (e.g. Edelman, 2008; Wieviorka and Calhoun, 2013) and discussing with academics close to La Vía Campesina[13] who had been through similar processes during their research. Such dialogues, combined with the insights from the social movement literature, taught me the importance of reflecting on my own approach and role as a 'researcher' and the importance of making explicit mutual expectations between the researcher and the 'researched'.

From the perspective of LVC, one secretariat staff support member explained to me in one of our first conversations that researchers seeking to interview social movement delegates during policy meetings may result in taking time away from the policy-processes, lobby work and networking.[14] These dialogues with the UN technical staff support meant that I was highly conscious of my own role when I participated in LVC meetings (see also Brem-Wilson, 2014).

In order not to overwhelm LVC delegates with interviews during global meetings, I conducted a considerable number of follow-up interviews when members returned to their villages thereafter. This methodology had several advantages. Firstly, it helped to compensate for the fact that I could not physically follow the members when they returned to their villages. Secondly, interviews conducted on Skype tended to be less hectic than interviews conducted during global meetings, where LVC members have to focus on political processes, networking, etc. Finally, interviewing activists in their own context – and not only during global meetings – meant that members were often more comfortable and in a better position to reflect on their own engagement and their participation in UN processes.

d Scholar–activist relationships: potentials and challenges

Over the last few years, the literature on research with and on social movements has considerably increased (Brem-Wilson, 2014: 111). In particular, we see an increasing integration of social movements' interests into academic processes, and relations of knowledge production are getting more and more attention from scholars (Wieviorka and Calhoun, 2013; Edelman, 2009; Brem-Wilson, 2014; Gillan and Pickerill et al., 2012; Pleyers, 2015). The increased interest in sharing and producing knowledge in favour of movements' struggles can be seen, for instance, in the launch of the online journal *Interface* with a mission to provide 'material that can be used in concrete ways by movements'.[15] Another academic project that attempts to build convergences around knowledge production by learning from and with social movements is the online platform *OpenMovements*.[16]

An increasing number of 'activist researchers' are working closely with social movements and are engaged in 'solidarity research;' i.e. producing knowledge that is relevant to their struggle (including critiques), but which does not harm or undermine them (Duncan, 2014: 78). Enhanced collaborations between movements and academics seems potentially significant for the incorporation of activists' worldviews into the academic analysis of movements, not just as objects of future studies, but as agents of emergent perspectives (McMichael, 2010).

Such constellations help to raise greater possibilities for both activists and academics to engage in meaning-making (Kurzman, 2011) and the production of alternative world-views that may challenge those hegemonic meanings that are often taken for granted (Duncan, 2014: 78; McMichael, 2010). Consequently, research on knowledge production can contribute to mutual benefits; activities and academics can support common interests and possibly enhance social transformation towards a more just and equitable society.[17] From the side of academia, examples of co-production of knowledge may be important so as to move from the ivory tower of academia towards more dialogue and more horizontal foundations for knowledge-building (Santos, 2007).

Scholar–activist relationships entail both potentials and challenges. For instance, Kurzman argues that movements who analyse themselves are in direct competition with academics outside of the movement, 'who may have a vested professional interest in downplaying activist knowledge-production or segregating this knowledge-production as an object of analysis that is distinct from their own scholarly acts of analysis' (2011: 11). Kurzman further contends that the constellations of movement–academic scholarship must not only be built on dual loyalties but also on clear demarcations of the different roles (Ibid.). In his study, Marc Edelman explores the synergies and tensions between rural social movements and professional researchers, portraying how movements' practices and strategies are often complicated by the fact that most university researchers who study or accompany social movements are profoundly sympathetic to the activists' goals (Edelman, 2008: 49). As movement activists seek to 'tell their story' they may directly or indirectly shape or determine the research agenda.[18] Consequently, Edelman insists that the contributions to agrarian movements may depend on the ability of the researcher to pose 'difficult questions' (Ibid.).[19]

As more 'activist research'[20] and spaces for sharing social movements' outlooks are gaining ground, several scholars have called for the need to carefully and critically scrutinise social movements–academic scholar relations (e.g. Touraine, 1981; Wieviorka and Calhoun, 2013; Edelman, 2009; Escobar, 2008). Edelman argues that the boundaries between activists and academics are not always as sharp as it is sometimes claimed.[21] The importance of keeping an appropriate degree of distance between movement activists and academics has, for instance, been expressed by Michel Wieviorka and Craig Calhoun in the 'Manifeste pour les sciences sociales' (2013). The authors contend that researchers struggling to keep enough distance from social movement actors constantly run the risk of being 'symbiotic' i.e. simply accompanying the actors, supporting an ideology, or identifying with them to a degree that it becomes unclear whether the researcher is a producer of knowledge, an actor, or an activist him/herself. It remains, however, the authors argue, important to build the valuable links between researchers and social movements as these in many cases can help to heighten the consciousness of actors themselves and the context in which they operate. This, the authors conclude, also contributes to heighten social movements' capacity for action (Wieviorka and Calhoun, 2013: 5).

By arguing that observers and research produced by social scholars contribute to social movements' self-reflection and agenda, I do not, however, suggest that social movement activists do not possess this capacity themselves. One of the remarkable aspects of my research with members of La Vía Campesina has been these individuals' willingness and openness to critically discussing their roles in the movement.[22] This interest in engaging in discussions with others to learn from critical dialogue reveals an awareness of the constant need for reflection and openness to be questioned by others. Such observations further affirm that those living in poverty are capable of reflection, expression and analysis (Touraine, 2007: 492; McMichael, 2010) – also at the global scale.

3 Conclusion

In this opening chapter of the book, I presented the approach and method to study LVC as actor and arena (Borras and Franco, 2009). The focus on agency and some of the challenges related to applying multi-site and multi scale research design was introduced. The chapter presented how the research evolved and unfolded some of the methodological considerations related to 'getting access' to a social movement. In particular, the chapter discussed the importance of establishing trust while being an 'outsider' in an 'insider' location (Dwyer and Buckle, 2009: 58). Based on my own experience of researching close to LVC members, this chapter concludes with a discussion of the potentials and challenges for building social movements–scholar relationships. In the concluding section of this book, I will return to how such new constellations may lead to more horizontal foundations for knowledge-building.

Notes

1 Borras and Franco state that it is this dual character – as both a single actor and an arena of action – that has made Vía Campesina an important 'institution' for national and local peasant movements, and an interesting but complex entity for

other transnational social movements, NGO networks and international agencies to comprehend (Borras and Franco, 2009: 11).

2 Or as Borras has phrased it: LVC is more than 'the total sum' of the different agendas and goals of its member associations (2004: 99).

3 A number of social movement scholars have drawn our attention to the social movement's culture, its imaginaries, meanings and significance (e.g. Alvarez et al., 1998; McDonald, 2006; Pleyers, 2010).

4 According to Touraine a social movement is a heuristic concept and not an empiric reality 'A social movement is neither an empirical reality, nor a transcended reality. It is a sociological concept. Historic actor can neither be fully identified to it, nor understood outside of their relation to it' (Touraine cited in Pleyers, 2015: 119). Pleyers adds that this approach is complementary to quantitative methodologies and the study of international institutions and policies.

5 From now on agency is to mean 'the capacity and will to act'.

6 As James Jasper has noticed in presenting a strategic approach to collective action: 'the subject of strategy has been almost entirely colonized by game theory' (2004: 3).

7 MST: Movimento dos Trabalhadores Rurais Sem Terra (Brazil's Landless Worker's Movement).

8 Examples of invited speakers during LVC's 6th International Jakarta Conference: Regional International Fund for Agricultural Development (IFAD) and FAO representatives, as well the UN Special Rapporteur to Food, Olivier De Schutter. (Mr Schutter presented a Skype message since he was unable to attend the conference in person.)

9 For instance, Jun Borras, Eric Holt-Giménez, Marc Edelman and Annette Desmarais.

10 See Duncan (2014: 83) for a discussion on participant observation and the importance of sensitivity while conducting research with civil society actors in the UN arena.

11 I was not fully aware of this sensitivity before a social movement activist explained to me the importance of being very careful while interviewing, in particular elderly persons who are not always comfortable with a very 'direct' style of interviewing.

12 For instance, it was required that I was careful in not sharing information with other researchers in the CFS who were not 'entrusted' by LVC or other social movements.

13 I am particularly grateful for the discussion I had with Josh Brem-Wilson, Priscilla Claeys and Tea Fromholt regarding this challenge.

14 Such concerns that researchers will disturb the internal processes partly explain while LVC (or rather secretariat staff/the International Council of LVC) accepts to engage with only a few researchers. This proves to be a good understanding of how social movements work. This was communicated to me in discussions with LVC leaders during the Jakarta Conference, June 2013.

15 http://www.interfacejournal.net/who-we-are/mission-statement

16 This platform was launched in 2015 by the International Sociological Association Research Committee on Social movements. https://www.opendemocracy.net/breno-bringel-geoffrey-pleyers/openmovements-social-movements-global-outlooks-and-public-sociologist

17 For instance, Santos emphasises the need for academics to 'interact politically with a world whose realities of social exclusion and inequality demand a proactive role' (Santos, cited in Brem-Wilson, 2014: 116).

18 Other scholars bringing critical reflection and practice that challenges, in diverse and sometimes contradictory ways, the scholar movement boundary are Touraine, 1981; Flacks, 2004; Routledge, 2004; Croteau, 2005; Escobar, 2008.

19 In my research questionnaire seeking to explore how and why social movement activists use their precious energy in formal inter-state institutions like the CFS, I

asked some sensitive questions, for instance, concerning the risk of co-optation and de-radicalization.

20 Other examples of related types of research: 'Public Sociology' (Burawoy, 2000) or 'queer public sociology' (Santos, 2012), 'politically engaged ethnography' (Juris and Khasnabish, 2013), 'militant ethnograph' (Juris, 2007), 'Participatory Action Research' (Kapoor and Jordan, 2009).

21 For instance the complex challenges facing today's social movements have required some activists to become researchers (Edelman, 2008: 247). Examples of leading LVC figures (or, in some cases, former participants) who have written perceptive 'insider' histories of the movement and published in academic journals are: Bové, 2001; Bové and Dufour, 2001; Borras, 2004, 2008; Desmarais, 2002, 2007.

22 Not only did LVC members show strong willingness and openness to discuss with me during global meetings and field visits, I also experienced how members of the LVC subsequently called me on Skype to hear more about the progress and results of my research.

2 The consolidation of an international peasant movement

Building on existing literature, this chapter explores the rise and development of La Vía Campesina as an international peasant movement (Desmarais, 2007; Borras, 2004; Borras and Franco, 2009). The first section explores how farmers' organisations came together in LVC as a global movement and united around one clear opponent: the dominance of neo-liberal globalisation as an expression of global capitalism (Desmarais, 2007: ch. 2). The second section sheds light on the power struggle within global civil society. It presents how LVC has been carving out a space and voice for itself in the international arena as a response to the tendency of others talking in the name of peasants (Desmarais, 2007: 23; Borras, 2008). Thirdly, this chapter examines the organisational functioning of LVC and argues that the movement is building another model that is different to the one predicted by the 'classical path', expecting social movements to develop from loose networks toward increased top-down structures, centralisation and bureaucratisation (Della Porta and Diani, 1999; Everett, 1992). In contrast to this model, this chapter shows how LVC seeks to organise around a model that maintains a low degree of institutionalization. It delves into its internal functioning and demonstrates that it is highly dependent on its members' ability to adhere to strong principles, such as participation, consultation and trust.

1 A counter-response to neoliberal globalisation

The following section presents how LVC has emerged, how its unites small and medium size farmers across the North–South divide (Smith, 2002), and increasingly articulates the peasant struggle under a vision for food sovereignty

a Peasants globalising the struggle

LVC emerged in 1993 as a concrete expression of peasants' concerns with the consequences of neoliberal globalisation. Peasants felt that they must come together at the global level to unite their forces against the growing global agri-business model's increasingly negative effects on small-scale food producers and their environments (Rosset and Martínez, 2014; Desmarais, 2007; McMichael, 2006). Amid the rise of neoliberal policies in the 1980/1990s, regional peasants and peasants' movements across the North and the South found ways to meet and exchange experiences, organise strategies[1] and analyse what was happening in the countryside, worldwide (Desmarais, 2007; Borras, 2010). These encounters enabled peasants and farm workers from around the world to find common ground in their struggle against a worsening crisis. In 1993, 46 members of national and regional peasants and farmers' movements from around the world gathered in Mons, Belgium, to formally constitute LVC. Here, they declared they would strengthen their ties and to forge international links between their organisations and other farmers' organisations from around the world. Instead of competing against each other, they decided to unite their forces. As declared in the Mons Declaration:

> Peasants in the North and in the South united around the opposition to agricultural trade liberalisation. We observe a deterioration in the overall situation in the rural world, expressed by deepening poverty across the whole planet and the massive exodus from the countryside, which is raising global unemployment levels and urbanising huge rural populations, in turn provoking a worsening of the desertification process with serious human and ecological consequences. Currently, hunger in a substantial part of the world is allowed to coexist with a situation of surpluses of all kind of agricultural products. This contradiction can only be understood as a result of the neoliberal agricultural and development policy promoted by government and International Organisations. It is at the same time, the main obstacle for peace and development for the peoples in the world.
>
> (LVC, 1993: 1)

As Nettie Wiebe from the Canadian National Farmers Union (NFU), founding member of LVC, testifies:

> I think we were aware that the forces behind the neoliberal model could not be challenged by us fighting on our farms alone. We decided to build a movement around strong relations towards land, seeds, nature and each other.
>
> (Canadian farmer and founding member of LVC, Nettie Wiebe interview, 30.01.14)

LVC often states in its documents that the key decision to set up LVC in 1993 was a reaction to the Uruguay Round of the General Agreement on Tariffs and Trade (GATT) – leading to the creation of the WTO. The Bangalore Declaration from La Vía Campesina's 3rd International Conference in 2000 asserts:

> We are united in our commitment to confront and defeat the global agenda of neoliberalism. The negative impacts of globalisation are acute and tragic in the countryside. The imposition of the World Trade

> Organization (WTO) and regional trade agreements is destroying our livelihoods, our cultures and the natural environment. We cannot, and we will not, tolerate the injustice and destruction these policies are causing. Our struggle is historic, dynamic and uncompromising.
>
> (LVC, 2000)

b Food sovereignty as a banner of struggle

Touraine argues that it is within a conflicting relationship with the adversary that a social movement constitutes itself (Touraine, 1978). The constitutive relationship with neoliberalism is manifested, as LVC builds its model around opposing sets of values and competing worldviews and development models (Desmarais, 2007: 33; Rosset and Martínez, 2010). LVC emphasises how the struggle is not a struggle between farmers in the North and peasants in the South but a struggle to radically transform the dominant agriculture and development model (Desmarais, 2007: 33). As the Maputo Declaration from LVC's 4th International Conference in 2008 states:

> La Vía Campesina formed in the North and South around common objectives: an explicit rejection of the neo-liberal model of rural development, an outright refusal to be excluded from agricultural policy development and a fierce determination not to be 'disappeared' and a commitment to work together to empower a peasant voice. Through its strategy of 'building unity within diversity' and its concept of food sovereignty, peasant and farmers' organisations around the world are working together to ensure the well-being of rural communities.
>
> (LVC, 2008: 41)

Whereas in the 1980s and the 1990s, scholars mostly focused on how civil society organised around single issues such as defending the environment and women's interests, LVC strives to presents holistic alternatives to neoliberal globalisation (Borras and Franco, 2009; Rosset and Martínez, 2014; Desmarais, 2007: 24). LVC presents food sovereignty as a political framework by integrating social, ecological, moral and ethical principles into a holistic vision to replace the ideology of neoliberal globalisation (Desmarais, 2007). Food sovereignty was first launched at the 1996 World Food Summit and has since been further developed. The shorter version of the Declaration of the Forum for Food Sovereignty, which was held in Nyéléni, Mali, 2007, provides a description of food sovereignty:

> Food sovereignty is the right of peoples to healthy and culturally appropriate food produced through ecologically sound and sustainable methods, and their right to define their own food and agriculture systems. It puts the aspirations and needs of those who produce, distribute and consume food at the heart of food systems and policies rather than the demands of markets

> and corporations. It defends the interests and inclusion of the next generation. It offers a strategy to resist and dismantle the current corporate trade and food regime, and directions for food, farming, pastoral and fisheries systems determined by local producers and users. Food sovereignty prioritises local and national economies and markets and empowers peasant and family farmer-driven agriculture, artisanal – fishing, pastoralist-led grazing, and food production, distribution and consumption based on environmental, social and economic sustainability. Food sovereignty promotes transparent trade that guarantees just incomes to all peoples as well as the rights of consumers to control their food and nutrition. It ensures that the rights to use and manage lands, territories, waters, seeds, livestock and biodiversity are in the hands of those of us who produce food. Food sovereignty implies new social relations free of oppression and inequality between men and women, peoples, racial groups, social and economic classes and generations.
>
> (LVC, 2007.[2] Full text in Appendix 3)

As this declaration shows, food sovereignty goes well beyond ensuring that people have enough food to meet their physical needs. The vision of food sovereignty stresses the importance of which type of food is produced, how it is produced and on what scale (Desmarais, 2007: 34). Unlike large monocultural plantations, small farmers often plant several different species of staple crops, such as potatoes, improving the resilience of their food and increasing its diversity. The declaration asserts that people must reclaim their power in the food system by rebuilding the relationships between people and the land, and between food providers and those who eat. Food sovereignty is thus a coherent vision placing autonomy, human relationships, dignity, well-being and diversity above a model of large-scale, capitalist and export-based agriculture (LVC, 2012c). McMichael argues that food sovereignty can be seen as a response to a 'top-down' neoliberal model where 'the market has become the dominant lens through which development is viewed' (2010: 3).

LVC presents food sovereignty as a vision for society that fulfils the human right to food and food security, but entails a more resilient development model to meet the challenges of the interrelated food, energy and environmental crisis (Claeys, 2013). LVC presents food sovereignty as a resilient model for society anchored in values such as solidarity, environmental responsibility, and the valorising of local knowledge and traditions (LVC 2015).[3] A European LVC member expressed during a Side-Event at the 41st Annual Session of the CFS, how food sovereignty is a frame that can help to build the core-principles of society:

> Food sovereignty and agroecology are coherent value practices that we can implement. We have the choice to build the world we want. We must always ask ourselves: do we wish an individualist world where people care for themselves and destroy nature and exploit workers? Or a world where

citizens engage in community struggles, care for the planet, each other and the next generations?

(European LVC member, CFS41 event on agroecology organised by the International Planning Committee for Food Sovereignty, Rome, 13.10.14)

Members of LVC often state that food sovereignty has been practiced for decades. As one settlement farmer explained when I visited his settlement farm in Santa Maria, Southern Brazil:

> Food sovereignty is not new. My grandparents were creating food sovereignty. Food sovereignty from the past allows us to procure healthy food for the next generations.
>
> (Settlement Farmer of the Assentamento Carlos Marighella, Santa Maria, Rio Grande de Sul, interview, 21.03.2013)

c Building support for food sovereignty across multiple scales

While peasants claim to have practised food sovereignty for generations, food sovereignty has increasingly become the banner of a struggle that allows for the incorporation of various agendas (e.g. agrarian reform, seeds and agroecology). As Claeys documents, food sovereignty can be seen as the mobilising 'rights master frame'[4] for LVC as it guides their positions, actions, and ways of being as a movement (2013: 48).

Members of LVC increasingly gather around the principles of food sovereignty, although this has not been easy for the movement, given the fact that it is constituted by a diverse member base with a multitude of visions on how to attain an overall frame of action. In her account of LVC's development years of consolidation, Desmarais demonstrates how the road to building unity around a collective vision of food sovereignty within LVC has been 'a long and industrious process' (2007: 33).

Today, LVC is actively building food sovereignty as a cultural and political framework to a radical extension of the ideas surrounding food security (Desmarais, 2007; McMichael, 2010; Witmann, 2009; Mann, 2014; Müller, 2013). At the national level, an increasing number of governments are integrating the concept into their national constitutions (Claeys, 2013: 352; McKeon, 2015: 153; De Schutter, 2013a). Over the last decade LVC has played a key role in advocating the vision of food sovereignty from within global institutions such as the FAO, the United Nations Commission on Human Rights Council (Desmarais, 2007; Claeys, 2013), and more recently within the reformed CFS (McKeon, 2015; Duncan, 2014). While it has proven to be an arduous struggle to bring the term food sovereignty into UN intergovernmental texts (McKeon, 2015), policies for food sovereignty policies are nevertheless being developed (see Chapter 4, this book).

Even though the term food sovereignty has not been recognised within a UN framework, a main task for LVC seems to be to demand policies that correspond to the major food sovereignty principles. This includes the notion of food as a human right, the recognition of small-scale food producers, more direct democracy and greater citizen participation in framing policies for food and agriculture.

Defining a collective overall frame of action that allows for multiple agendas and issues to be explored has not only been important to hold members of LVC together, it has also been instrumental in developing a broader framework that can work for a large global movement that aims to challenge the dominant neoliberal model of food production and consumption (Desmarais, 2007: 42; McMichael, 2009). While divergent views and tensions remain, the concept of food sovereignty is increasingly deployed to challenge the dominant model of industrial agriculture and to shift towards a model that protects, prioritizes and strengthens small food producers and local food systems. This fight for people's food sovereignty is part of the larger paradigmatic struggle over the global food system (McMichael, 2006; Desmarais, 2007; McKeon, 2015). It is against what McMichael (2005) calls the dominant 'corporate food regime' and 'modernity's promotion of monoculture' that will lead us to 'agriculture without farmers' (McMichael, 2014: 17) that LVC fiercely seeks to re-shape our development model in order to bring production and decision-making back into the hands of people (Mann, 2014; Desmarais, 2007).

2 Carving out spaces for peasants in the international arena

a Building a transnational collective peasant identity

The existence of a transnational social movement depends on its ability to develop a common interpretation of reality, and to nurture solidarity and collective identifications (Touraine, 2001; McAdam et al., 1996: 8). Without a sense of identity, there can be no 'real struggle', as Freire puts it in *Pedagogy of the Oppressed* (1970).

From the early beginnings of the movement in Mons in 1993, members of La Vía Campesina have strived to establish and define their identity as 'people of the land' (Desmarais, 2008: 139). This collective place-bound identity is particularly important for the internal dimension of a peasant movement uniting peasants across borders, cultures and class differences[5] (Borras, 2004: 9). According to Touraine (2001), the establishment of a transnational social movement requires the development of a discourse that identifies both a common identity ('the us') and the target of protest ('the other') at a supranational level (Touraine, 2001). As explained above, LVC was born in direct opposition to the dominance of neoliberal globalisation and in particular to the increasing threats of the growing agribusiness industry. Rosset and Martínez have expressed this opposing relationship in their study of the evolution of LVC as a transnational movement in the following way:

> In the neoliberal era, supranational corporations and institutions dictating neoliberal policies have negatively affected most sectors of society. One of

> the consequences of this is that class or cultural differences are no longer the barrier they once were for transnational collective action. In fact, rural organizations and peasantries around the world share the same global problems even though they confront different local and national realities.
>
> (Rosset and Martínez, 2010: 150)

Canadian farmer and founding member of LVC Nettie Wiebe confirms that the strong perception of a clearly defined opponent (Touraine, 1978) has been indispensable to building a transnational collective peasant identity:

> I think that as we realised the severity of the corporate model all over the world we became more and more convinced about the need to unite. We shared a vision for the restructuring of agriculture and food sector that still unites us more strongly. With the multi-crisis, we see even more clearly the consequences of the neoliberal model and importance of uniting against it.
>
> (Nettie Wiebe, Skype interview, 30.01.2014)

Fieldwork conducted with members of LVC in different regions of the world shows that members often (however not always) use the term 'peasant' to describe themselves (Edelman, 2008; Desmarais, 2007). Whereas the term 'peasant' is commonly associated with resistance to process or even backwardness, LVC and other rural movements are reclaiming *the political significance* of the word when they speak of the 'People of the land' (Desmarais, 2007; Borras 2004, 2008a). While LVC members use different and sometimes overlapping terms to describe themselves, the term 'peasant' represents a recognition of and respect for autonomy, control over land and territories, rural lifestyles as well as a rejection of large-scale industrial agriculture (LVC, 2014d; Mann, 2014; McMichael, 2010). More importantly, when members of LVC come together under the peasant category, it is not an attempt to flatten their diversity, but rather a process of coming together around a political culture and common struggle. As expressed by Basque farmer, Paul Nicholson:

> It is not a question of seeking out differences in order to synthesise or explain them. It is that the word 'campesino' has a clear meaning in the Latin American context and the word 'peasant' has a reality in the context of India, for example, that it doesn't have in England or the United States or Canada. We call our movement La Vía Campesina, understanding that it is a process of peasant culture, a peasant 'way'. The debate isn't in the word 'farmer' or 'peasant'. The debate is much more about the process of cohesion.
>
> (Paul Nicholson. Interview conducted by Hannah Wittman cited in Patel, 2009: 678)

Despite the ongoing debate over definitions, LVC is a movement that to a large degree unites around the political significance of 'peasants' as a way to partially reclaim the very meaning of peasant in a modern world (Desmarais, 2007: 197; Borras, 2004: McMichael, 2010; Edelman, 1999).

As Chapter 4 will further explore, the political significance of the peasant identity is important in the attempt to spell out the different roles and responsibilities while mobilised peasants today engage in the power struggle and attempt to shape the terms of the debate within the CFS.

The following section sheds light on LVC's struggle to carve out a space for peasant farmers in the international arena, historically dominated by NGOs and big farmers' associations, notably the International Federation of Agricultural Producers (IFAP).

b Two main rivals in the international arena

LVC is not the first organisation that gathers farmers' at a global level. LVC was founded upon a desire to be a movement that represents the values and demands of marginalised peasants in the countryside and in debates on food and agriculture (Edelman, 2008; Desmarais, 2007; Borras and Franco, 2009). According to Edelman (2008) and Desmarais (2002), the then dominant organisation known as the International Federation of Agricultural Producers (IFAP)[6] – which was founded in 1946 – was mainly interested in supporting larger farmers but was speaking in international arenas and negotiations as if it were representing *all* farmers.

Desmarais elucidates how LVC resisted the tendency among international bodies and institutions to 'lump'[7] rural actors into one category; one single 'farmers' space' that dilutes their differences (McKeon, 2009a: 29; Desmarais, 2007: 88–103). In a similar vein, Nettie Wiebe, from the National Farmers Union of Canada, explains how IFAP, after the foundation of LVC in 1993, has attempted to co-opt the movement to legitimise its own positions on several occasions in international meetings and 'spaces', by seeking to erase the fundamental differences between the two international farmers' organisations:

> In international meetings we were expected to unite around one statement only and to talk with one voice. In such meetings IFAP forcefully strove to undermine and to silence the proposals of La Vía Campesina.
>
> (Nettie Wiebe, interview, Skype, 30.01.2013)

We will further examine how this power struggle between peasant farmers and agribusiness farmers plays out within the CFS as a political battlefield in Chapter 4.

Another historical rival of LVC in the international arena are non-governmental organisations (NGOs). NGOs have historically claimed to be the mediators or the 'transmission belt' (Steffek et al., 2008: 3) between ordinary citizens and institutions of global governance, to help marginalised and mute actors to find an effective voice (Desmarais, 2007: 21–24; McKeon, 2009a: 9; Tilly, 2004: 152). The tendency of NGOs to 'monopolise the microphones' (McKeon,

2009a: 9) culminated with the UN summits in the late 1990s. In 2006, this process led Basque farmer and founding member of LVC, Paul Nicholson, to argue that:

> Peasants no longer want NGOs to be speaking in our name to international institutions and about agricultural policy. We want to build our own movement, our own international network and to speak for ourselves.
>
> (Bamako, Policycentric World Social Forum, quoted in Pleyers, 2012: 177)

LVC is clearly seeking to carve out a 'distinct and direct voice' for small- and medium-scale farmers (Borras and Franco, 2009: 28; Desmarais, 2007: 21–24) but this must be seen against the backdrop of its problematic context. There are attempts to lump all non-governmental organisations together, which leads to the blurring of significant differences, unequal power relations and conflicts that generally constitute the global civil society sphere (Desmarais, 2007: 88–103; Kaldor, 2003: 84).

In Chapter 7 I return to the ongoing tensions between social movements and NGOs in the context of both players working together in the same civil society 'interface' with the CFS. Here, LVC engages in a permanent struggle to reveal the unequal power relations, different mandates and values held by the different constituencies. The following section presents how LVC has organised itself as a 'movement of movements' and, to a large degree, united itself around a shared peasant identity and common values.

3 Key organisation principles

a Demanding accountability from 'the bottom up'

From its very beginning, LVC has strived to build a movement around principles of consultation and accountability (Desmarais, 2007: 28; Borras, 2004). As we will see in the following section, the struggles for the strongly held values of autonomy, inclusion and participation can be seen as an attempt to build an organisation model based on a low degree of institutionalisation (cf. Table 2.1). LVC as a peasant organisation can be categorised as a people's organisation (PO), where leaders are accountable to a social constituency and where representatives are appointed to different positions by, and accountable to, a member base (Desmarais, 2007: 23; Borras, 2008). In its attempt to build an organisational model with a strong focus on a social constituency and on principles of accountability, the movement opposes itself to 'autocratic and unaccountable policy-making' (Patel, 2009: 669).

Below, Nettie Wiebe explains how the rationale of LVC is clearly a counter-proposal to 'top-down policies' and to 'neo-liberal efficiency imposed from above'. Additionally, another Vía Campesina representative offers interesting observations about the importance of horizontal organising.

> In La Vía Campesina we are critical of the emphasis on 'efficiency' as a key determinant or value. It results in organisational monoculture – which is very efficacious but it is not resilient, it is not creative and it is damaging. Hierarchical structures are so effective in organisation that they overcome the need for trust – in a power structure managers can send their employees down the road, if they want. In this sense it is efficient for time management and very top-down. This is not how we work or think in La Vía Campesina. La Vía Campesina is built around autonomous organisations that are linked together. We don't have that concentration of power. The alternative to the clear top-down power-structure is to build trust. This is a lateral way of working and it is built on interchange and respect. This relies on healthy processes and builds strength. It is more effective for us.
>
> (Nettie Wiebe, interview, Skype, 30.01.2013)

In La Vía Campesina the horizontal structure means that decisions take time. This is, however, considered to be more effective in the long run. As expressed by an food activist from a farmer organisation, applying for membership of La Vía Campesina:

> Hierarchal top-down structures can be very efficient if speed is the only consideration, but they often fail further down the track due to a lack of consultation, passive resistance, exclusion etc. The horizontal structure and functioning of LVC is not as organisationally efficient as top down – many things take time as consultation with all constituents is mandatory – but it is far more effective in the long term. It is effective because bringing the people with you is as important (probably more important) than the objective. In fact, I would say that it is impossible to achieve food sovereignty from the top down. This means being organisationally efficient rather than effective, which can often be even counterproductive.
>
> (interview, Skype, 04.12.2014)

b Principles of decentralisation and autonomy

In her in-depth account of LVC's internal dynamics, Desmarais describes how LVC seeks to build a model where power and control are in the hands of members of the movement from around the world (Desmarais, 2007: 21). Given LVC's strongly held principles of decentralisation and autonomy, the movement's model of decision-making is rooted in the decentralisation of power in nine regions of the world (LVC, 2015,[8] see also Desmarais, 2007: ch.1). The coordination among the regions is taken up by the International Coordinating Committee (ICC), which is composed of one woman and one man for every region, elected by the member organisations in the respective regions. The ICC meets twice a year to engage in collective analysis of global agricultural issues, to define joint action and advocacy at the international level, and assess compliance with the International Conference agreements. LVC also has a number of International Working Commissions focusing on different issues. A special feature of LVC is the absence of a top-down sovereign authority, dictating what any member should do (Patel, 2009: 143). The International Conference, which takes place every four or five years, has been declared as the highest decision-making body of the movement (LVC, 2014a). Here, delegations from all regions engage in collective analysis and policy development, as well as negotiation and consensus-building processes (Rosset and Martínez, 2010: 164).

The responsibility of the day-to-day organisation of the movement rests on the principle of a non-policy-making International Operational Secretariat, which coordinates actions and implements the agreements reached at the movement's International Conferences. It is worth highlighting that the International Operational Secretariat rotates between LVC's organisations and regions every four to eight years, which also illustrates how LVC seeks to develop a model based on a low degree of institutionalisation. Finally, another feature that demonstrates this is the fact that the operational secretariat is assisted by a limited number of staff employees (especially when compared to the size of the movement), who are located in the different regions.

c Values of participation and inclusion

The movement's functioning at its internal meetings/spaces reveals its strongly held values of participation and inclusivity. The movement's political culture is clearly anchored in these values. For instance, during the 6th International Jakarta Conference, I recall looking around the plenary room and observing peasants sitting around the room. When they left their seats to speak, they lined up in front of the microphone, carefully respecting the gender principle of their position in the line; alternating between one man and one woman. During the Jakarta meeting, one African LVC member told me that patience can derive from a strong will to mobilise your attention and learn from others, even though meetings can sometimes seem cumbersome:

> Meetings can at times feel endless. We talk and talk and sometimes we do not even know if we will come to an agreement. But our movement is not a political party that needs to reach an agreement at the end of the day. We keep talking until we find an agreement. If we don't agree we continue the following day.
>
> (African LVC member, interview, Jakarta, 10.10.2013)

In his study of activists in the 'Alter-globalization movement', Pleyers discovered that there is 'a price to pay for low institutionalisation' (2010: 177). For instance, experimentation with more inclusive processes, such as open mass-meetings, can lead to somewhat unstructured, draining discussions that do not lead to any concrete results.

Other social movement scholars have explored and presentenced different experiences with participatory democracy. For instance, in *Freedom Is an Endless Meeting: Democracy in American Social Movements*, Francesca Poletta (2002) challenges the conventional wisdom that participatory democracy is worthy in purpose but unworkable in practice.

One African LVC member, participating at LVC's International Conference in Jakarta 2013, explained the meaning and benefits of working in accordance with strong values of inclusiveness:

> The effort you made to keep patience is often rewarded with a certain satisfaction of being able to work together. Learning to listen does not mean being weak, on the contrary, it means being aware of one's limitations and mistakes, being willing to learn from others. Patience is an important value. You get a feeling that you become stronger this way. At the end of the day you become stronger by learning to work together united through different cultures and experiences.
>
> (African LVC member, personal discussion, Jakarta, Indonesia, 13.06. 2014)

Observations of and conversations with LVC members reveal how members of the movement held values on participation and inclusion. The strong determination to unite in diversity is important for LVC as a pluralistic movement working together. This was emphasised, among others, by the Basque farmer, Paul Nicholson, in an interview after the Jakarta conference:

> In La Vía Campesina we seek to turn what others see as a weakness, a high level of inclusiveness, into strength. For instance, when people leave the room after a long discussion, where all had the chance to present their views in the processes towards consensus, the ownership of the final result is normally very strong. These are strong moments of strong feelings of building a movement together.
>
> (Paul Nicholson, interview, Skype 30.06.2013)

Members often emphasise the importance of *process* and the advantages of learning from other cultures and experiences and express a strong will to arrive at common strategies. When Canadian participant Nettie Wiebe read the final collective draft of the 'Jakarta call' out loud during the last day of the Conference, the room was filled with a strong sense of hope: that it is possible – and of utmost importance – to find common ground and move forward in diversity.

Seeking to enhance inclusive structures and active participation, LVC weaves a net of horizontal 'spaces of dialogues'[9] comprised of manifold exchanges (Leff, 2004: 15–24; see also Rosset and Martínez, 2014). Building a space for debate where strategies and tensions can be expressed helps members to create a better understanding of other members' realities. This is an important element of the internal functioning, which helps the movement to unite in diversity (Desmarais, 2007: 75).

4 Between structure and network

One of the most debated issues in social movement research is the extent to which formal bureaucratisation helps or hurts movements (Piven and Cloward, 1977; Jasper, 2004; Pleyers, 2010). For instance, Gamson ([1975] 1990) found that groups were more likely to be effective (e.g. by getting media attention) if they had formal and well-organised structures. Piven and Cloward (1977), in their study of the poor people's movements, notably segments of the working class in the United States, argued that social movements become less contentious once they build formalised organisations. Other scholars have shown how democratic and egalitarian structures and network-based organisation can be more effective in favouring tactical innovations and collective creativity and be better adaptated to fast-changing environments among social movements (Polettta, 2002; Castells, 1996; Pleyers, 2010: 78).

The definition of which organisation model is the most effective for any given movement depends on the movement's goal. For LVC, the rejection of neoliberalism, patriarchy and the very model of capitalism on the one hand, and the members' distribution of power and the building of an inclusive movement structure from 'the bottom-up' on the other hand are central to the raison d'être of the movement.

As presented in the previous section, LVC's organisational model is built around a social constituency where leaders are entrusted to make decisions on behalf of their members in the regions and are immediately accountable to their membership or constituency. This way of organising, based on democratically elected leaders, is particularly important to make sure that the interests at the local level are reflected in the dynamics of the regions and at the international level (Desmarais, 2007: 23). LVC avoids bureaucratic structures, which have often been associated with unions, corporations and the state (Offe, 1987), and thus strives to build a model that combines the advantages of being a flexible network while recognising some degree of delegation/responsibility structure. As one African LVC representative put it in an interview during the LVC's 6th International Conference in Jakarta:

> We need some kind of leadership. In our movement we have leadership in the ICC. These are our entrusted leaders elected by members in the regions … Our movement may look vertical from the outside as we delegate to the ICC. But for us it is important to have a structure. It is important for the functioning of such as huge movement that there is some defined responsibility.
>
> (African LVC member, interview, Jakarta, 13.06. 2014)

Scholars have shown that a pure network culture and lack of responsibility may also create new layers of bureaucracy (Freeman, 1972). For instance, Freeman notes in *Tyranny of Structurelessness* that paradoxically, 'in the desire to avoid specialisation and task division and making everyone responsible for everything', this ultimately means that no one is accountable for anything, and spaces can rapidly either fall apart as essential tasks remain undone or those with the most resources and/or commitment take on more work (Freeman, 1972: 1).

LVC attempts to find a way between the entirely loose network structures – defined by Hardt and Negri (2005: 217) as 'distributed network', – and the 'professional leadership' in more bureaucratic hierarchies that has haunted other movements, such as the international labour movement (Vieira, 2011) and the 'old left' (Rosset and Martínez, 2010: 156).[10] To this end, LVC's organisational model can be described as a 'structured network' that strives to be 'tight but flat'. This was shared with me in an interview with Nettie Wiebe, reflecting on the everyday running of LVC as a global movement:

> Being a movement that is lateral and at the same time organised and tight – you need to have a tremendous amount of trust – this is not a commodity you purchase on the market. It is like seeds, you select them carefully and you take care. This is one of the characteristics we try to nurture, trust at all levels of the organisation.
>
> (Nettie Wiebe, Skype interview, 30.01.2014)

Trust-building seems to be a particularly important factor to the internal working of the LVC. This is particularly relevant when the movement engages in a complex international policy terrain. Here members often refer to how 'entrusted leadership' is followed by responsibility. Trust seems to be important to help the movement evolve.

As Passy (2003) argues, building strong ties is vital for large and flexible coalitions where trust and interpersonal bonds must be created to meet some of the challenges related to the delegation. Furthermore, Bennett (2003) found that a high degree of trust, which is important for social movements' internal well-functioning, is not commonly found in larger, more bureaucratic organisations. Chapter 6 will explore how the peasant delegates to the CFS interviewed for this work express a strong *will* to act as 'entrusted leaders' when they are engaged in global policy work.

a A global rural movement in a digital age

In seeking to build an organisational model with inclusive structures, La Vía Campesina has been aided by the fact that it has consolidated in a time of rapid spread of new information communication technologies. The geographical expansion of new technologies is important for LVC to build itself as a transnational agrarian movement and to connect between farmers organised in rural areas.[11]

Increased access to Internet and the speed of new technologies provide the last generations of movements born in a 'digital age' (Castells, 2000; Passy, 2003) with options to build global-level strategies that go way beyond those of their predecessors. For instance, free online tools like multi-lingual Skype calls and Google Translate can help facilitate protest manifestations taking place at the same time in different places (e.g. the International Day of Peasants' Struggle, 17 April). This empowers the movement to better shape its public image (Mann, 2014: 154–162). An active use of communication technologies is instrumental in sending out regular online press releases, presenting video reportages about the struggle to a broader audience, building alternative media, and strengthening ties to friends and allies. New communication tools help social movements expand their networks. Thus they can be more efficient when seeking to build a different model of internationalisation to the ones predicted by the most predominant theories (Tilly, 2004; Tarrow, 1998).

The following section sheds light on the debate related to funding strategies, in particular the importance of diverse sources of funding, resilience and autonomy in order to build the argument that does not automatically follow the 'classical path' of internationalisation that is often connected with strong institutionalisation (Tarrow, 1998; Tilly, 2004).

b Autonomy and the issue of funding

A common view expressed by social movement scholars is that social movements – for the sake of their survival – will end up relying on more resourceful actors and transform themselves into 'respectable NGOs' (Kaldor, 2003; Kriesi, 1996).

Whereas NGOs often tend to look for funding opportunities from as many agencies as possible, LVC is highly selective of agencies to approach, or from which agencies to receive funding (Borras and Franco, 2009: 29). LVC is funded by the contributions of its members, by private donations, and by the financial support of some carefully selected NGOs, foundations, agencies, and local and national authorities.

As it strives to retain a high degree of autonomy, LVC wishes to avoid being too dependent on only one source and has developed a funding strategy that focuses on diversity.[12] It has been well documented how the flow of official aid money has 'distorted' the work of autonomous civil society by making it subject to government agendas and control (e.g. Brehm et al., 2004). While financial means are needed to facilitate the work of a global social movement, there is no doubt that money from private companies and foundations risks jeopardising the autonomy of the movement. As an activist expressed in an interview during a CFS session in 2014:

> All systems, LVC included, are only as resilient as the depth and breadth of the diversity within the system. The more diversity in breadth and depth, the more resilient the system/organisation. This is an absolute truth and it doesn't matter if the system is biological or a human construct. So if funding resilience (and organisational resilience against outside interference) is an aim or goal, then diversity of funding sources is an absolute necessity.
>
> (Activist from an organisation applying for membership of LVC, Skype interview, 01.12.2014)

Additionally, members of LVC often raise concerns that bureaucratisation of fundraising processes may have a demobilising effect on members. This also explains why LVC members seek to pursue alternative funding strategies. An example of this is that LVC has started training programmes to elaborate financial strategies and develop funding methods as a part of its strengthening of the membership at the regional level (LVC technical staff support person, Skype interview, 01.07.2014).

During LVC's 6th International Conference in Jakarta, a pending member of LVC expressed his excitement that family farmers had found ways to go beyond official channels, and use such strategies as crowd-funding to increase funding for their participation. Drawing from his own experience, he explained how farmers' sharing of the practices and working together can be a way to 'sidestep' bureaucracy:

> We can make things happen in a few days because we avoid bureaucracy. This is why we sometimes manage to move faster than political leaders and institutions. Moreover, we achieve more with much less money or resources. This means that when we do get small amounts of diverse funding we get a lot done for the money.
>
> (Pending member to LVC, interview, Rome, 18.10.2013)[13]

While some farmers in some regions are innovative in finding alternative sources of funding, not all members of LVC may have the same capacity – or the same possible sources available – to fundraise to support their own participation to global meetings.

For instance, one of the LVC support staff persons explains how the movements 'core-funding budget' funds the participation of members to take part in key international meetings. In every type of meeting the movement strives to ensure that regional representation is accomplished by having two people (one man and one woman) attend (LVC staff support person, Skype interview, 1.07.2015).

Apart from the core funding budget, LVC has a budget that is used for thematic meetings. This budget can be used, for instance, to support participation in other types of international meetings, such as participation in thematic meetings on seeds, investments in agriculture and other issues related to LVC's engagement in global meetings in the food security institutions in Rome[14] (LVC technical staff support person, Skype interview, 01.07.2015). These examples illustrate that the movement cares deeply about its autonomy, and is aware of the consequences of the various choices at hand. It seems ready to pursue alternative strategies should the old ways prove to jeopardize its autonomy.

5 Modes of organisation/operation

Table 2.1 is a synthesis that illustrates how institutionalisation could be conceptualised as a degree rather than an absolute category. The table summarises

Table 2.1 Continuum of modes of organisation/operation

Low degree of institutionalisation	*High degree of institutionalisation*
• Decentralised organisational structures, 'bottom-up' driven efficiency (consultation, inclusion)	• Hierarchical organisational structures, 'top-down' efficiency
• Process orientated, experimentation with participatory methods	• Result- and outcome-driven focus
• Leadership with a collective sense, principle of rotation, 'reflexive' leadership	• Tight vertical structure of delegation, hierarchical individualistic leadership
• In principle apolitical and light secretariats of few staff (compared to the size of the member base) • Locations often in the Global South • Demand for autonomous spaces	• Policy-making secretariats • Locations of head office often in the Global North
• Seeking autonomy in funding activities and diversity of sources	• Strong funding dependence on few sources
• High concern for internal democracy (internal accountability, representativity, legitimacy, transparency)	• Scant concern for internal democracy (internal accountability, representativity, legitimacy, transparency)

some of the organisational characteristics of institutionalisation crystallised from the social movement literature and the empirical data from this research. Rather than suggesting mutually excluding categories, the table is an *ideal type* (Max Weber [1922] 1995) that may be useful as an analytical tool to show that institutionalisation may be approached as a continuum/a degree.

Whereas some of the traits of a high level of institutionalization may be reproduced at different times and in different contexts, the point is to show that LVC does not automatically follow a pattern of institutionalisation that goes from low to high. One example illustrating that these categories are not mutually exclusive is how LVC is well aware that 'some degree of delegation is needed'.[15] This is evident in the movement's choice to organise as a peasant/people's organisation where LVC leaders delegated to global meetings are appointed to these roles by a member base and rural constituency, to whom they are accountable (Borras, 2008; Desmarais, 2007: 23). By remaining well aware that the different organisational/leadership dynamics are not rigid and mutually exclusive when studying the features of LVC as a global movement, it can be argued that the movement places itself predominantly on a low degree of institutionalisation.

6 Conclusion

While civil society internationalisation in the 1990s was characterised by a high degree of institutionalisation and a taming process that quickly transformed new social movements into respectable NGOs (Kaldor, 2003; Pleyers, 2012), LVC engaged in another process. This chapter has shown how the movement has built an organisational model centred on principles of decentralisation, collectivity, and inclusiveness. As an expression of a people's organisation, LVC's political culture reflects a low degree of institutionalisation, including strong principles of participation, consultation, accountability and representativity. By building this organisation model, the movement is opposing itself to the structures of dominating neoliberal institutions, and some NGOs (Patel, 2009: 669; Desmarais, 2007).

Before I examine some of the practical implications for LVC to retain a low degree of institutionalisation when moving into the structures of a UN arena, in the next chapter I explore how rural actors have played a key role in carving out an autonomous space for participation in the UN Committee on World Food Security.

Notes

1 For instance 'farmer-to-farmer' exchanges of agroecological methods that peasants have practiced for generations and are continuing to improve in context-specific ways (Holt-Giménez, 2006).
2 http://nyeleni.org/spip.php?article290
3 This cultural framework resonates with the indigenous people's cosmo-vision with a respect for nature, valuing the primacy of community over individualism; the well-being of people over profit (Pleyers, 2015: 109).
4 A rights master frame functions as a collective action frame, which has been identified as sufficiently broad in interpretive scope, inclusivity, flexibility and cultural resonance to function as master frame (Snow and Benford, 1988: 198)
5 Activist scholars (e.g. Borras 2004; Vieira, 2011) have identified the recognition of different ideological positions in the movement as a particular challenge for La Vía Campesina.
6 Since 2012 IFAP has been replaced by the World Farmers Organization (WFO) which is based in Rome.
7 Desmarais (2007: 119) presents a historical overview of how global actors in different ways have attempted to co-opt LVC to legitimise their own positions, emphasising the difficulties they faced in the Global Forum for Agricultural Research (GFAR).
8 http://Viacampesina.org/en/index.php/organisation-mainmenu-44/what-is-la-Via-campesina-mainmenu-45
9 'Diálogo de saberes' (Leff, 2004; Rosset and Martínez, 2014: 1) or 'assemble' (Escobar, 2008).
10 Rosset refers to the 'old left' as being built up around 'individualistic, personalistic, and clientelistic leadership' (2015:156).
11 For instance research conducted within the network of Farmers' and Agricultural Producers' Organisations of West Africa, Ouagadougou Burkina Faso 2010, revealed how mobile phones are an effective way of connecting farmers. Fieldwork research conducted with peasants in West Africa and Brazil affirms that direct phone calls are the most efficient way to connect with fellow peasants.
12 This was communicated to me by LVC staff member, Nico Verhagen (Skype interview, 01.07.2014). Also see Desmarais, 2007, who explains LVC's strategic building of alliances and selection of funding sources.
13 This food activist explained how a very new small band of volunteers produced a comprehensive national food policy document called the *The People's Food Plan:*

http://www.australianfoodsovereigntyalliance.org/peoples-food-plan/. This plan, based on food sovereignty, was, in fact, more complete than the Federal government's own plan. This plan was done with a 100% volunteering effort from farmers, academics and others, plus the costs were minimal (<AUD 800), in a period of 6 months – in contrast to the government's plan that cost Australian tax payers AUD 4.5 million and took 2 years. To make matters even more, the government changed and the new government cancelled the AUD 4.5 million plan (Member of the Australian food sovereignty alliance, pending member to LVC, interview, Rome 18.10.2013).

14 The Rome-based UN agencies are the following: the Food and Agriculture Organization of the United Nations (FAO), the International Fund for Agricultural Development (IFAD), the World Food Programme (WFP) and the Committee on World Food Security (CFS).

15 As expressed by An LVC member during the Jakarta conference, 12 June 2013.

3 Carving out an autonomous space for participation in the UN Committee on World Food Security

This chapter explores how LVC has played a key role in carving out an autonomous space for civil society participation in the Committee on World Food Security (CFS), employing a pattern that does not follow the classical path of internationalisation with institutionalisation. LVC and its allies have carved out an autonomous space for *diverse* civil participation in the CFS by establishing an 'International Food Security and Nutrition Civil Society Mechanism' (CSM).[1] This chapter reveals that the pattern built by social activists engaging in the CFS is neither linear nor static but rather dynamic, complex, and negotiated through relationships. Before presenting how the CSM organises itself around the principles of self-organisation and autonomy, while functioning as a global connecting mechanism, the chapter starts with a short overview of global civil society's interface with the UN. In this brief presentation of these historical processes, the idea is not to give a full account, but rather to shed some light on how LVC has played a central role in carving out the CSM, in particular from its parallel engagement in the International Planning Committee for Food Sovereignty since the World Food Summit in 1996.

1 From UN summits to autonomous spaces

Since its creation, the United Nations has promoted civil society participation in various processes of dialogue and deliberation (Cardoso, 2003; Willets, 2006).

In the 1970s and the 1980s an increasing number of activists from civil society organisations started to carry their concrete problems and struggles to UN arenas, culminating with a 'participatory turn' that accelerated in the 1990s after the end of the cold war,[2] when a proliferation of global networks were formed (McKeon, 2009a: 11). At this time, national-level NGOs started to emerge at the global level in greater numbers and sought to engage directly in inter-governmental deliberations and advocacy work. They were particularly active in the fields of human rights, environmental and gender issues (Bringel, 2015: 131).

These developments led to a series of World Summits and Conferences throughout the 1990s such as, among others, the World Summit for Children

1990, the UN Conference on Environment and Development 1992, the World Conference on Human Rights 1993, and the World Food Summit 1996. These meetings were sites where civil society actors could come together to engage in debate, put forward ideas and proposals, and take advantage of lobbying opportunities. The nature of these global meetings led to criticism from some social movement activists who stated that this kind of engagement was more 'window dressing' from the UN than a meaningful space for participation and change (McKeon, 2009a: 173). As McKeon observes in her assessment of what the author has called the 1990s' decade of UN summits: 'Civil society actors felt increasingly that they were in serious danger of being co-opted to serve watered down intergovernmental agendas than advancing their own visions and objectives' (McKeon, 2009a: 10–11).

Notably, given the predominance of neoliberal ideology as a guide to integration and global governance, many society organisations, even some of the NGOs that had initially invested energy and enthusiasm in the global UN processes, became increasingly sceptical of the United Nations' potential to address basic global problems (McKeon, 2015: 92–93 and 2009a: 10–11; see also Smith, 2008: ch. 9). Consequently, civil society actors increasingly turned their energies towards building alternative global fora rather than let their engagement with the UN be interpreted as legitimising UN outcomes they did not necessarily agree with (McKeon, 2009a: 9–11). The most well known in the literature may be the World Social Forums (WSF) and their regional offshoots, which were originally held as a people's counter-meetings to the business-driven World Economic Forum's Annual Meetings in Davos (Pleyers, 2012; Caruso, 2013).

Although LVC played a central role in the WSF processes in its early days (and even today still uses this platform to build and maintain alliances and networks) observers and some LVC activists have criticised the nature and questioned the legitimacy of the WSF (Pleyers, 2010; Vieira, 2011). This may explain why LVC has increasingly turned its energies towards building other types of alliances, deliberations and autonomous spaces around food sovereignty, while maintaining a strong focus on internal movement building (Borras and Franco, 2009: 38; Rosset and Martínez, 2010). As we will see in the following section, LVC has played a central role as a working member of the International Planning Committee for Food Sovereignty (IPC), which has given a strong impetus to global networking[3] and the opening up of a political space within the FAO.

2 World Food Summits and opening of political space at the FAO

The engagement of rural organisations with and within the FAO, in particular, the civil society forums held in parallel to the World Food Summits in 1996 and 2002, differed considerably from the earlier UN processes mentioned above. Unlike the NGO-dominated civil society meetings that traditionally accompanied official UN summits, rural movements increasingly played a central role in the parallel events to the FAO World Food Summits as part of the International Planning Committee for Food Sovereignty (IPC). The IPC network emerged ahead of the 2002 'World Food Summit. Five years later'. The network describes itself as an 'autonomous and self-organised global platform of small-scale food producers and rural workers' organisations and grass root/community based social movements to advance the food sovereignty agenda at the global and regional level'.[4]

The IPC has particularly invested considerable energy to gain recognition and rights for civil society actors to self-organise in relation to the area of food and agriculture, and in particular with and within the FAO (McKeon, 2009a: 50–120; McKeon and Kalafatic, 2009: 17–18). A central achievement of the IPC has been to systematically increase the presence and voices of rural actors in different ways.

For instance, the IPC helped to increase the participation of peasants and indigenous peoples in global meetings by setting up a quota system and by mobilising resources for their travel. This helped to ensure that small-scale food producers and indigenous peoples were in the majority (McKeon, 2009a: 22). The IPC also played a key role in the global network that organised around the International Conference on Agrarian Reform and Rural Development (ICARRD) that was held by the FAO in March 2006 in Brazil. During the ICARRD Conference, the IPC served as the official anchor of the NGO parallel forum attended by approximately 500 civil society representatives from different regions of the world.

This organising process facilitated by the IPC allowed the issues of grassroots groups linked to land, such as peasants but also pastoralists, nomads, indigenous peoples, and subsistence fisher folk, to be loudly and systematically brought to the surface (Borras and Franco, 2009: 25).

3 A reform process with strong emphasis from below

The experiences around the IPC have been indispensable for the establishment of an autonomous civil society mechanism (CSM) to the CFS. The IPC's almost two decades of experience in opening up spaces for peoples' organisations at the FAO meant that it was well positioned – both in terms of capacity but also in terms of political recognition by FAO – when the food crisis reached the media headlines in late 2007 and the reform of the CFS gained traction (McKeon, 2015: 95; Brem-Wilson, 2011).

Among those engaged in the CFS related work,[5] one often hears that the CSM is the 'son or daughter of the IPC'. As one European civil society activist close to the CFS reform put it in an interview: 'Without the contribution of a well-organised people's movement like La Vía Campesina, ready to invest energy here, it would never have been happened' (European civil society activist engaged in the 2009 CFS reform process, interview, Rome, 14.10.2014).

After eight months of intergovernmental negotiations, the proposal to reform the CFS into an inclusive international and intergovernmental platform – for which civil society activists in the network of IPC had successfully lobbied – defeated other proposals (McKeon, 2009; Brem-Wilson, 2011; Duncan, 2014). One LVC member engaged in the CFS reform explained in an interview that the actual pace and scope of reform moved much faster than civil society representatives had expected. For instance, as stated by a European LVC member who followed the CFS reform process from its beginning: "I did not expect the processes would go so fast. I think most of us, including many government representatives, were not really aware of how far we got here" (European LVC member, interview, Rome, 13.10.2013).

The 2009 CFS reform showed how UN reform can be accomplished with strong impetus from below.[6] The following section presents the composition and some key points of the reformed CFS, followed by an examination of the functioning of the autonomous Civil Society Mechanism (CSM).

4 The composition and function of the post-2009 reformed CFS

The CFS is hosted in the FAO Headquarters in Rome with a joint Secretariat composed by FAO, IFAD and WFP. The renewed CFS structure consists of a CFS plenary session held annually, where the member governments and other CFS stakeholders gather to make decisions, debate, and coordinate on food security and nutrition issues. Negotiations on the development of key documents take place throughout the year in the Open-Ended Working Groups (OEWGs). Besides that, the CFS Bureau is the executive arm of the CFS, which is supported by an Advisory Group (AG) supporting the Bureau to advance the objectives of the CFS. The Plenary of the Committee on Food Security is held annually in October and reports to the UN General Assembly through the United Nations Economic and Social Council (ECOSOC).

The CFS reform document differentiates between 'members' (UN member states), 'participants' and 'observers.' Participants are divided into five categories:

(I) Representatives of UN agencies and bodies with a specific mandate in the field of food security and nutrition;
(II) Civil society and non-governmental organisations and their networks with strong relevance to the issues of food security and nutrition;[7]
(III) International agricultural research systems;
(IV) International and regional Financial Institutions, regional development banks, and the World Trade Organization (WTO);
(V) Representatives of private sector associations and private philanthropic foundations active in the areas of concern to the Committee.

(CFS, 2009: Rev.2 paragraphs 2–4)

Box 3.1 presents some of the key points of the CFS 2009 reform document.

Box 3.1 Committee on World Food Security: key points of the 2009 reform document

- Recognizes the structural causes of the food crisis and that the primary victims are small-scale food producers.
- Defines the CFS as 'the foremost inclusive international and inter-governmental platform' for food security. Includes defending the right to adequate food in its mission.
- Explicitly includes defending the right to adequate food in the CFS's mission.
- Recognizes civil society organizations – small-scale food producers and urban movements especially – as full participants, for the first time in UN history. Authorizes them to intervene in debate on the same footing as governments and affirms their right to autonomously self-organize to relate to the CFS through a Civil Society Mechanism.
- Enjoins the CFS to negotiate and adopt a Global Strategic Frame-work (GSF) for a food strategy providing guidance for national food security action plans as well as agricultural investment and trade regulations
- Empowers the CFS to take decisions on key food policy issues, and promotes accountability by governments and other actors through an 'innovative' monitoring mechanism.
- Enjoins the CFS to adopt a Global Strategic Framework (GSF) for food strategy providing guidance for national food security action plans and other CFS participants and as a reference point for coordination and accountability.
- Arranges for CFS policy work to be supported by a High-Level Panel of Experts in which the expertise of farmers, indigenous peoples and practitioners is acknowledged alongside that of academics and researchers.
- Recognizes the principle of subsidiarity and promotes linkages between the global meetings and multi-stakeholder policy spaces at regional and country level. Governments commit to establishing national multi-stakeholder policy spaces in the image of the global CFS.

(CFS (2009) summarized by McKeon (2015: 108). Reproduced with permission.)

While the CFS describes itself as a multi-stakeholder platform that 'enables all viewpoints to be considered when deciding on concrete actions to address issues affecting food security and nutrition'[8] the term 'multi-stakeholder platform' is highly contested. Research conducted in the CFS arena shows that many civil society activists engaged in the CFS reform reject the use of the term 'stakeholder' as it blurs the essential differences between those engaged. As Duncan wrote 'It is argued that the stakes of peasant farmers and multinational corporations are not at same, and to call them both stakeholder depoliticizes the important difference and differentiation of power' (Duncan, 2014: 91).

Many social movement actors believe that by categorising everybody as 'stakeholders', the CFS runs the inherent risk of diluting states' responsibilities and of sustaining existing power imbalances (Observation, plenary discussion during the CSM Annual Forum, Rome, 10.10.2015). In this regard, the methodology of the CSM has been fundamental for the vision of social movements to spell out the identities, interests, roles, and responsibilities of different actors engaged in the governance processes (McKeon, 2015: 199). In Chapter 4 we return to how these struggles play out when LVC engaged in the reformed CFS.

Finally, one component of revitalising the Committee on World Food Security in the 2009 reform process has been the establishment of a High Level Panel of Experts (HLPE). The role of the HLPE is to provide background information and analyse and advise the CFS with reports on various issues upon request of the CFS.[9] At the heart of the vision of the HLPE as scientific panel is the recognition of expertise beyond academia and scientists. HLPE presents itself as an effort 'that should help create synergies between world class academic/scientific knowledge, field experience, knowledge from social actors and practical application in various settings' (HLPE, 2010). In the context of the contested debate of what 'evidence-based' policy-making means, the declared mandate of the HLPE has been described as a victory for civil society (McKeon, 2015: 107). The following section presents how the autonomous Civil Society Mechanism was built with a strong impetus from 'below' as a space for civil society engagement for the CFS.

5 The autonomous Civil Society Mechanism as an interface to the CFS

For LVC and its allies, one of the key achievements in the CFS reform processes has been the establishment of an autonomous Civil Society Mechanism (CSM) as a recognised channel for autonomous civil society input to CFS deliberations (CFS, 2010). The CSM presents itself as a mechanism that

> facilitates dialogue between CSOs in all continents, sharing information with them on global policy debates and processes, promoting civil society consultations and dialogue, supporting national and regional advocacy and facilitating the participation of a diverse range of CSOs at the global level of policy making processes.[10]

The CSM is made up of a general membership organised around 11 constituencies (smallholder farmers, fisher folk, pastoralists, landless, urban poor, agricultural and food workers, women, youth, consumers, indigenous peoples, and NGOs) from 17 sub-regions, a Coordination Committee, Working Groups, and a Secretariat. The Coordination Committee consists of 41 members selected through processes established by representatives of the constituencies or sub-regions. Through the CSM, CSOs have engaged in many aspects of the work of the CFS, both as members of an Advisory Group, Open Ending Working Groups and the yearly plenary sessions, where civil society and social movements are acknowledged as full participants (CFS, 2009).

The CSM secretariat is deliberately 'lightweight' and staffed by only three people. The secretariat is, in principle, non-political and plays a facilitating role in the engagement of the CSM to the CFS. This organisation model with a deliberate 'light' secretariat means that the success and operational functioning of the CSM secretariat depends on the active participation of its members, resource people, and particularly, the Coordination Committee members (CSM, 2014a). The interaction between the CSM and the CFS is illustrated in Figure 3.1.

a The 2009 CFS reform: opening up the box of civil society

Scholars continuously disagree on how to define civil society. For instance, it remains an ongoing debate whether or not to include corporations or other for-profit actors or non-political associations in the definition of civil society (see e.g. Gramsci, 1971; Marx, 1847; Habermas, 1997; Scholte, 2001; Wapner, 1996). In this regard, the CFS is a platform for political engagement that

Figure 3.1 How the CSM engages with the CFS
Source: CSM, 2013

recognises the need for negotiation among diverse actors within non-governmental actors and civil society itself. This was made explicit during the CFS reform process, where non-governmental actors were called to organise autonomously in order to facilitate their interaction and engagement with the Committee. This led to the creation of two separate and self-defined mechanisms: a Civil Society Mechanism (CSM) working separately from the Private Sector Mechanism (PSM).

The design[11] of an autonomous civil society mechanism that works separately from the private sector[12] has been particularly important for social movements to reveal the power struggle and the different roles and mandates within civil social society.

As Chapter 2 showed, recognising the different roles and mandates within global civil society has been part of LVC's historical struggle to carve out a space for itself in the global arena and in doing so the movement has revealed the different visions, identities, roles, interests and power struggles that exist within global civil society actors working on agriculture and food issues (Desmarais, 2007: 21–24). In this historical context it is remarkable that the CFS acknowledges the different constituencies and clarifies the importance of different types of contribution that each ought to make. As stated within the section of 'participation of civil society/Non-Governmental organisations':

> It is important to recognize that civil society cannot be considered a single homogenous voice. Different kinds of organizations carry out different functions, defend different interests, and can lay claim to different forms of legitimacy. The distinction between non-profit CSOs and private sector associations is a fundamental one. *So is the distinction between NGOs which undertake advocacy, emergency or development activities in favour of the rural populations of developing regions, on the one hand, and rural peoples' organizations encompassing farmers' and rural producers' organisations which have a mandate to speak on behalf of their membership and are accountable to them, on the other.* It is important to ensure that there is effective input from all the different constituencies and regions.
>
> (FAO, 2008: Paragraph 22, my emphasis in italics)

By emphasising the different mandates and memberships between these organisations, the CFS can be theorised to consider civil society as including the spaces of social life that lie outside states and economic interests: what the CFS calls 'the private sector'. This is close to Habermas's approach of making a distinction between civil society and economic actors (Habermas, 1997: 394–399).[13]

In the example of the CFS, the methodology for civil society participation differs from other UN meetings where civil society spaces have been lumped together in groups, effectively ignoring their different mandates, identities and power struggles (Desmarais, 2007: 21–26; McKeon, 2009a: 56–57). In particular, granting civil society members the right to design their own engagement in an autonomous civil society mechanism is unique: 'The way the CFS was designated makes it more complex than other inter-governmental and other types of stakeholder fora. This is why the CFS is more interesting for us to invest our time and energy here' (European LVC member, interview, Rome, 20.10. 2014).

b CSM diversity versus UN consensus

Ahead of CFS plenary sessions, members of CSM seek to find points of agreement to put forward united positions and statements. This includes, among others, agreement on the so-called 'red lines' or 'bottom lines' of what the group is willing to accept on the most contentious issues. These lines are discussed and decided on during the CSM Civil Society Forums that are held at the FAO ahead of the CFS Annual Sessions (Field notes, CFS-rai negotiations, Rome, May 2014).

A CSM secretariat member explained in an interview that the secretariat seeks to facilitate consensus positions between the members whenever possible, however, the CSM secretariat does 'not push for consensus' (CSM secretariat member, interview 10.04.2013). The CSM accepts diversity, difference and disagreement and this is clearly stated in the organising principles of the mechanism: 'The CSM presents common positions to policy makers where they emerge and the range of different positions where there is no consensus'.[14] A CSM member participating in designing the mechanism explained this logic of consensus:

> Consensus agreements are based on achieving unity beyond differences. The road to consensus is a long process, but we take the time that is needed. We sit down and we discuss until we find a consensus or decide on what to do.
>
> (European food activist, member of CSM, interview, Rome, 10.10.2013)

In occasions where consensus is reached among civil society constituencies, it means finding unison among a wide variety of viewpoints and sometimes very different positions. Peasant activists, who emphasised their diversity, have repeated this point. They underscored that their positions in the CSM are not reducible to one voice. However, members are also strategically aware of the political weight they can gain in the CFS by uniting their positions: 'When we unite our voices we are stronger, and it is more difficult for governments to ignore our positions' (African LVC leader, CFS40 Annual session, interview Rome, 16.10.2013). Thus, while strong differences may exist between members of established groups, it can sometimes be advantageous for the group to temporarily unite around common positions. This is what Spivak (1988) calls actors 'essentialising' themselves in order to bring forward their group identity in a simplified way to achieve certain goals.[15] In this regard, the CSM members often reiterate that the 'CSM consensus', built on strong dialogues, process and respect for diversity, is very different from the 'UN consensus'. I return to this permanent tension within the CFS arena in Chapter 5.

c A voice – not a vote

In the CFS, all stakeholders are allowed to engage on an equal footing in procedures and debates leading up to final decision-making in the Committee. However, the CFS final voting authority remains in the hands of governments (CFS, 2010). It is important to emphasise that LVC and other CSM actors do not want a vote in the CFS. During the reform process, civil society was against the extension of voting rights beyond states for mainly two reasons: a) to guard against the dilution of the principle of member states' responsibility for hunger elimination, and b) to prevent the door from being opened to multinational corporations gaining voting power.[16]

One LVC member engaged in the CFS reform process explains that this methodology is important for the activists in the CSM to 'have their hands free' to hold governments accountable for their decisions afterwards. The attempt to exercise political influence without acquiring political power is in line with Habermas's definition of civil society presented above (Habermas, 1997: 394). Civil society actors interviewed for this book often state that their engagement in the CFS is very different from engaging in UN forums and conferences during the 1990s. As one European LVC member put it when reflecting on the CFS five years after the reform:

> The CFS is a very special UN Committee. It is more complex, which makes it a more interesting model for us. We are not made co-responsible for the outcomes. We engage to the last minute and then make governments decide. If governments do not listen to our suggestions, it remains their full responsibility that people continue to be hungry.
>
> (LVC member, interview, Rome 10.18.2014)

The fact that governments maintain the monopoly to vote is important for LVC to guard against institutionalisation in the traditional sense. This was described as follows by one of the LVC leaders who had been engaged in designing the methodology for civil society engagement to the CFS:

> When we engage here it does not mean that we give up our local or global strategy in LVC. This strategy remains. We do not vote and do not adapt our own strategy to the UN. However, we participate at the CFS as maybe the UN can give us some tools to strengthen our struggles. This may sometimes happen, sometimes. What is important is that in the CFS we retain our own strategy when we engage in these global processes.
>
> (European LVC member, interview, Rome, 20.10.2013)

As the LVC member elaborates, this methodology for civil society engagement in the CFS is noticeably different from other UN multi-stakeholder models:

> It is government consensus and governments' products that is coming out of the CFS. We use these documents when we can, but it is not our

> documents. We keep our hands free and we always follow our own strategies.
>
> (European LVC member, interview, Rome, 20.10.2013)

The engagement in CFS means that civil society representatives must be very good 'equilibrists'. This delicate balance was expressed by a European LVC member as follows:

> Here [in the CFS] consensus means that you agree to participate, to engage your own personal time and resource to strategise, to struggle…and then you don't vote. It can be a very difficult challenge for some members.
>
> (European LVC member, interview, Rome, 20.10.2013)

d Key principles of autonomy and self-organisation

In the beginning of this chapter, I presented how the autonomous Civil Society Mechanism (CSM) is an example of civil society actors carving out an autonomous 'interface' mechanism (McKeon, 2009a; McKeon and Kalafatic, 2009) to engage within a UN Committee. For LVC in particular, the possibility of social movements to play a central role in designing the CSM around principles of autonomy and self-organisation has been crucial for the movement to engage in this UN Committee. A Latin American-LVC member expressed this is at a CSM evaluation in 2014. The farmer put it like this: 'In the CSM, the terms of engagement have been politically negotiated terms. It is unique as we have our own space from where we define our own rules and terms of engagement' (Latin American-LVC member, CSM evaluation of the CFS Annual Session, Rome, 15.10.2014).

In interviews, LVC leaders often emphasised the importance that the CSM remain an autonomous space, in the sense that it does not become subject to governments agendas and control. In this regard, building a mechanism seeking to guard against external actors 'meddling' into the internal affairs of the CSM seems to be central for civil society actors when they decide to engage within the UN intergovernmental world (McKeon, 2009a: 141–157). In this sense, inventing an autonomous space for civil society engagement to the CFS can be understood as what Jonathan Fox terms as: 'the ability of citizens to organise in defence of their identities and interests without fear of external intervention and punishment' (Fox, 1994: 151–152).[17]

For others to truly understand and respect social movements' autonomy is, however, an ongoing struggle. One LVC member engaged in the CFS processes stated after a CFS meeting, how he felt it was tiresome that members of CSM group often needed to remind governments that autonomy means both time and resources for civil society actors to organise themselves autonomously:

> Autonomy here is somehow different from our local struggles. We must remember that we do not have 100 percent autonomy. Here we have a set of rules to follow – and we still need to learn how to navigate what

> autonomy means at this level. It a constant struggle. For example, if we do not have money for translation to participate fully – how can we be autonomous here? We are only one actor in this landscape and we need to struggle for our autonomy at this level, also. We need resources and time to work autonomously.
>
> (Latin American LVC member, interview, 15.10.2013)

As the LVC member expresses here, autonomy is not a 'gift' but in practice a constant struggle to achieve. This affirms that autonomy is relative and not absolute (Borras, 2004). The example also shows how autonomy as an antagonistic concept is 'a site of struggle' (Bohm et al., 2010: 28). The innovation and dynamics around building the autonomous civil society mechanisms bring nuances to the relationship between participation and autonomy as well as the claim often put forward by scholars: that inclusion automatically leads to a loss of autonomy. For instance, Woolford and Ratner contend 'the price of political inclusion for social movements is the loss of their autonomy – the very quality that provides them the space necessary for a creative re-imagining of the social world' (cited in Smith and Dawn, 2008: 206). In a similar vein of thought, post-development scholar Escobar warns that engagement with the state and other political institutions is a threat to the autonomy of local cultures and potentially undermines the solidarity of resistance (Escobar, 1995).

In contrast to these predictions, the CSM autonomous space, according to members within the group, not only helps activists to self-organise, reflect, co-build and evaluate their positions, but also this engagement helps to build solidarity among members.

> There are so many spaces of neoliberalism. The CSM is our space. We built it. Here we can learn from each other and build our positions. We all become stronger when we are building bonds of solidarity with each other. This strengthens us here and in our struggles back home.
>
> (European LVC member, CSM meeting, evaluation, 10.10.2014)

In this context, the CSM shows the innovation and political imagination of social actors finding ways to adapt their claim of autonomy to a global political context and win autonomy vis-à-vis the institution. The example shows that there is not necessarily a contradiction between participation in a political institution and retaining a considerable degree of autonomy.

6 NGO – social movements: collaborations and tensions

a Bringing those most affected into the leadership

At the core of the CSM is the understanding that those most affected by food insecurity and malnutrition must be agents of their own development and change. As the CSM Annual Report (2012) stated:

> The goal of the CSM is to facilitate participation from broad sectors of society with a wide range of views by giving priority to those most affected by food insecurity and malnutrition to ensure that policy decisions are being influences from the ground-up.
>
> (CSM, 2012: 5)

In particular, the CSM calls for special attention to the organisations representing the various constituencies of the food insecurity (CSM, 2010a: paragraph 10). These constituencies are numerically dominant in the governance of the mechanism. As stated in the proposal to establish an autonomous civil society mechanics to relate with the CFS:

> One of the key organising principles is for self organised groups to speak for themselves in the CSM and have a greater representation in the mechanism; of these self organised constituencies, smallholder producers have a larger number of spaces in the coordination mechanism because they represent the majority of the world's hungry; they also hold in large parts solutions to addressing hunger sustainably. Whilst recognising and affirming the role of self organised constituencies, the CSM will ensure that issues and voices of those who are unable to organise find space within the CSM.
>
> (CFS, 2010a: 3, note 2)

One important pillar of the CSM is to distinguish between the roles of the political leadership (food insecurity, social movements) and those with a role of technical support (NGOs, academics). One example of the CSM seeking to fulfil its vision is that the CSM is designed to favour the engagement of peoples' organisations, with only one out of 11 constituencies for NGO delegates (CSM, 2013).

Yet, intrinsic advantages and power imbalances remain in the CSM as NGOs often have more resources and experience to engage in lobby and advocacy work in UN arenas. Thus, even though the CSM in its vison and design pays special attention to the participation of social movements and others who may be most affected by food insecurity, there is a constant risk that NGOs will end up filling the political leadership position.

Field observations in the CSM space show that the challenges related to this tension between social movements and NGOs is constantly addressed among the members. For instance, in one of the daily CSM preparatory meetings during the Annual CFS40, one of the CSM secretariat members reminded the group that it was not possible to make any political decisions without social movements in the room (Field observation, CSM meeting, CFS Annual Session, Rome, 16.10.2014). The reason that no social movements were present in this preparatory meeting was that there were parallel lobby meetings with governments going on in other rooms at the same time.

Reflecting on the challenges related to LVC's engagement in the CFS five years after the reform, Paul Nicholson, Basque farmer and founding member of

LVC, explains that the resource constraints mean that LVC sometimes needs to select 'entrusted political support people' to participate in meetings when social movements cannot be present (Paul Nicholson, Skype interview, 30.10.2014). Data from fieldwork in the CSM affirm that this does occur. For instance, in an internal CSM meeting one LVC member expressed the importance of delegation and trust building between social movements and NGOs in the following way: 'We need you here. We cannot be in Rome all the time. We must trust you as our sisters and brothers' (LVC member, CSM meeting ahead of Annual CFS40, 11.10.2015). Those NGOs that may be delegated to be facilitator in the CSM meetings or working groups are often individuals who are conscious of the political responsibility entrusted to them by other members. As reported by Duncan[18] citing a NGO actor coordinating a working group in 2012:

> It is not the role of the NGO, as an NGO person, it shouldn't be my role and it's not what I wanted to do. But, also, I think that there was sufficient trust between each other and there was also kind of, I think a relatively good, how do you say that… reporting back and preparation and reporting back what the issues were … I also felt that there was sufficient backing in terms of the positions that we defended, so it wasn't really a difficulty. There was kind of a mandate by a broader group.
>
> (Representative of a large International NGO, Rome, interview conducted by Duncan, June 2012)

The example shows how trust seems to be important to cultivate this kind of unique space and collaborative action. Field data collected from following the CSM working group on investments (2012–2014) affirm that the tensions between social movements and NGOs that are often prevalent in Coordination Committee processes tend to dissolve or at least soften in policy working groups (Duncan, 2014: 146). In Chapter 7 I further examine some of the current challenges related to the ongoing tensions between NGOs and social movements in the CSM. The following section explores the specific mandate of the CSM policy working groups.

b The CSM policy working groups

The aim of establishing policy working groups on a year-round basis is to provide a space for CSOs to discuss specific issues to ensure civil society members can frame the debate in between sessions, increase awareness and share information on related CFS processes with the broader member base of the CSM (McKeon, 2015; Duncan, 2014). The strategy and the positions of the CSM that are planned during the year in the CSM policy working groups are finally discussed in the yearly CSM Civil Society Forum. While CSM strategies are discussed during the annual CSM Civil Society Forum, members of the CSM may develop their own strategic policy framework in the thematic policy working groups.[19] For instance, during the negotiations the CFS-rai members of the CSM working group on investments established their own negotiation document containing the autonomous reflections and standpoint of the group. This document included a full range of alternative wording. As it can be difficult to foresee how negotiations will play out, it is crucial for the group to have different strategy scenarios ready for the CFS. The CSM policy working groups can be seen as collaborative spaces[20] where the group develops joint positions to present at the CFS.

Observations in the CSM working group on investments showed how working groups are spaces for discussion, reflection, analysis and 'learning by doing'. For instance, in May 2014, ahead of the first round of negotiations on the principles for responsible agricultural investments, I remember finding myself in a preparatory meeting of the CSM working group on investments, where civil society actors actively strove to identify and address when 'responsible investments' violate international human rights. During this day alone, the group went through 68 pages of drafting text and a pile of juridical human rights declarations.

The dedicated efforts of the different civil society members to understand the legal aspects of the negotiations and their preparation well ahead of the CFS session display a strong sense of responsibility of members of the group to educate themselves to do the job that they are delegated to do at the CFS. The collaborative engagement in preparatory working groups meetings reveals a steep learning curve when new members arrive to participate in the CSM. A young member of the indigenous peoples' constituency expressed how she felt her own agency strengthened after engaging in the CSM: 'I am not an expert in UN language and I did not study international politics or law. But collective wisdom is produced here. It is like going to a people's University' (Spokesperson for the indigenous peoples, newcomer to the CSM, interview, Rome, 15.05.2014). A few days later, this young indigenous peoples' leader was chosen to be the spokesperson presenting a message during the CFS negotiations. In an interview she proudly expressed that she had formulated her speech based on her own experience in Latin America. Members of the CSM group actively support each other and largely recognise that some are more experienced than others.

> As human beings we make mistakes. Sometimes you speak terribly in public or you forget a central point. We are human beings and this is a natural part of a learning process. We are stumbling on the way, but we are learning from our mistakes – we must seek to transform mistakes into constructive learning experiences that we can build on. And then we can come out even stronger.
>
> (Latin American, LVC member, interview, Rome, 20.10.2013)

These learning aspects of the CSM illustrate how the members of the mechanism stress the importance of developing a strong internal process, when they engage in the CFS.

c The importance of consultation and process

In their work within the CSM, LVC and its allies stress the importance of civil society autonomy, self-organising and a process that involves participation, evaluation, and learning. For example, a declared statement of the CSM is that it will 'actively seek out and support the engagement of those mostly affected by food security issues and provide opportunities to hear alternatives of those more connected to the realities on the ground' (CSM, 2010a).[21] During the CFS negotiations on the investments in agriculture and food systems (CFS-rai), an LVC member emphasised the importance of conducting autonomous consultations within the CSM's constituencies to include more effectively the voices of 'those on the ground' in the UN processes.

> Global negotiated tools cannot simply be the results of a technical consultation among people here in Rome. These should start from the reality of citizens around the world. The consultations rooted in the regions are important to achieve the goal of bringing here the voices of those most affected by food insecurity and malnutrition. While this is not always easy to reach out to all them we want to speak to, we need to continue to build our own ways of doing consultation. We cannot wait for governments to make this happen.
>
> (European LVC member, interview, Rome 10.10.2013)

A document circulated within the CSM ahead of the CFS-rai negotiations in 2014 describing the processes of the autonomous CSM consultations in the regions, also expressed this rationale.

> The decision to do autonomous consultations was made by the civil society group from the logic that it is a key concern to include the social base on the ground at an early stage to define what the consultation would be about and what the process would look like.
>
> (CSM, 2013)

A European LVC member explained how the autonomous consultations conducted by the CSM demonstrate that the 'the CSM takes its autonomy seriously' and seeks to ensure that action within the CSM is not limited by the set of tasks outlined and defined by the CFS (European LVC member, interview, Rome, 13.10.2013).

Social movements' representatives also emphasise that the Civil Society Mechanism retains the option to 'walk out' of the CFS plenary sessions if the members of the group find that the rules of the CFS process are not being respected. One European LVC member explained that there is 'no such thing as a golden rule' for the group to use the 'exit strategy' to walk out of the CFS in protest.[22] This, the LVC members explained, depends very much on how the specific CFS process goes.

> In the CFS we agree on the process, not on the outcome. We can sometimes live with a less satisfying outcome, but we cannot compromise on the process. If we lose the process we lose the CFS. And then we have no reason to be here.
>
> (European LVC member, interview, Rome, 13.10.2013)

The examples above show how LVC members seem to be concerned with the UN *process,* and in particular that the CFS respects the principles of inclusiveness and participation of social movements and civil society organisations. In Chapter 5, I return to some of the ongoing challenges for social movements seeking to uphold principles of low institutionalisation.

7 Conclusion

Whereas most theories of social movement institutionalisation predict a process towards homogenisation (Kriesi, 1996; Tilly, 2004), the process of civil society actors designing an autonomous civil society mechanism to engage with the CFS has led to increased diversity beyond the participation of those actors normally dominating the UN meetings (CSM, 2010a: paragraph 10). This is a step towards expanding the opportunities to hear alternative voices that are often more connected to the realities on the ground than professionalised NGOs. Notably, the institutional innovation of designing an autonomous civil society mechanism (separate from the private sector) from where civil society organisations can self-organise and engage in all stages of the policy processes, with voting remaining in the hands of governments, is remarkable. This design has been important for civil society to hold governments accountable for their decisions, and, if necessary, to express their disagreement with the final UN document.

Finally, whereas diversity has often been portrayed as a hindrance to efficiency (see e.g. Desmarais, 2007: 43), the CSM group strives to 'pool' together their different expertise and experiences to make diversity an advantage while engaging in the UN arena. In this regard, clarifying different roles and mandate, and building foundations for trust, seems to be particularly crucial for the overall functioning of the CSM.

In contrast to what most dominant theories of social movement institutionalisation have predicted, social movements' engagement within established institutions does not automatically lead them to being devoid of innovation and losing their autonomy. The next chapter explores some of the dynamics arising from social movements and civil society actors taking part in international negotiations within the reformed CFS.

Notes

1 From here on 'CSM' or 'the Civil Society Mechanism'.
2 Scholars have stated how the changing political context in the post-cold war era of globalisation helped to open up the space of international deliberations. According

to McKeon it led to a 'more visible and effective role to a wider variety of civil society organisations, as the contribution of non-state actors to solving world problems was increasingly recognised in a paradigm of structural adjustment and redefinition of public/private spheres and responsibilities' (McKeon, 2009a: 51).

3 Some of the milestones have been the Nyéléni Forum for Food Sovereignty, held in February 2007 in Sélingué, Mali and the People's Food Sovereignty Civil Society Forum, held in Rome, Italy, November 2009. The first international meeting on agroecology held in Nyéléni, Mali, in February 2015 can also be seen as a landmark in this regard.
4 http://www.foodsovereignty.org/about-us/
5 For instance, registered in conversations and interviews conducted in the FAO house in Rome (2013–2014) which hosts the CFS.
6 McKeon (2015) and Brem-Wilson (2011) has presented how the presence of organised civil society played a central role and helped to optimise the CFS reform opportunity.
7 The paragraph continues: 'with particular attention to organisations representing smallholder family farmers, artisanal fisher folk, herders/pastoralists, landless, urban poor, agricultural and food workers, women, youth, consumers, Indigenous Peoples, and International NGOs whose mandates and activities are concentrated in the areas of concern to the Committee. This group will aim to achieve gender and geographic balance in their representation.'
8 http://www.fao.org/cfs/cfs-home/en/
9 http://www.fao.org/cfs/cfs-hlpe/en/
10 http://www.csm4cfs.org/about_us-2/what_is_the_csm-1/
11 Ultimately, the CSM was designed 'to ensure the pride of place and visibility of organisations directly representing small-scale food producers and other categories of those most affected by food insecurity' (CSM, 2010).
12 The PSM is represented by the International Agri-Food Network, an informal coalition of international trade associations involved in the agri-food sector at the global level and created in 1996. The aim of the network is to facilitate informal liaison among the professional organisations and towards international organisations in the agri-food chain at global level (Duncan, 2014: 124).
13 According to Habermas, civil society functions on a benevolent basis.

> [It] mobilis[es] from alternative sources of knowledge' and enacts political influence, but does not strive to take political power. In order for civil society to exercise its influence in the public domain (space) of a democratic society three conditions are necessary. First, civil society should be able to act in the frame of a rationalised world, second, its actors should exercise political influence but not acquire political power, third, civil society has to be able to transform the political system indirectly by drawing on and mobilising alternative knowledge.
>
> (Habermas, 1997: 394–399)

14 http://www.csm4cfs.org/about_us-2/what_is_the_csm-1/
15 Strategic essentialism is a major concept in postcolonial theory and was introduced in the 1980s by the Indian literary critic and theorist Gayatri Chakravorty Spivak. It refers to a strategy that can be used by different nationalities, ethnic groups or minority groups to oppose against the levelling impact of global culture (see Ashcroft et al., 1998: 159–160).
16 This has been communicated to me by actors following the CFS reform processes, for instance in an exchange with Brem-Wilson, researching LVC's engagement in the CFS reform processes (2009–2010).

17 Other social movement scholars have theorised how 'spaces of autonomy' (McDonald, 2007: 49–63) and 'spaces of experience' (Pleyers, 2010: 43–46) have helped social movement activists to live along with their principles and develop various social relationships to express their subjectivity.
18 Duncan did her research on and offered technical support to the CSM in 2010–2012.
19 The thematic policy working groups are: land tenure, agriculture investment, global Strategic Framework, gender, nutrition, price volatility, protracted crisis and conflicts, monitoring and mapping, climate change, biofuels, food waste and losses, water, and the CFS programme of work (CSM, 2015).
20 Term used by an African LVC member, CFS session, Rome, 10.10.2013.
21 http://www.csm4cfs.org/about_us-2/what_is_the_csm-1/
22 Data from observations in CSM meetings show that there are different opinions among civil society actors concerning whether the group should ever use walking out as an option. For instance, a NGO member of the CSM group communicated during a CSM meeting to other members in the group, that the group must be aware of the fact that it can be 'a dangerous game' for civil society to walk out of the CFS during a session. The NGO representative here referred to the risk that this action may be in the interest of those actors in the CFS wishing to strengthen the image of the CSM group as being non-cooperative. In addition, the NGO representative warned that walking out of the CFS Plenary may risk weakening the governance of the CFS, so that governments use less time and energy in the CFS and increasingly turn to less democratic governance constellations of food such as the new G8 Alliance for food security and nutrition (NGO member, CSM preparations, CFS 40th Annual Session, Rome, 16.10.2013). The walkout strategy has, however, been used by civil society actors in CFS. For instance, Brem-Wilson presents how a civil society walkout took place during the 2012 Plenary on the Roundtable on Price Volatility in response to their exclusion by the Chair in this session (Brem-Wilson, 2015: 16; see also Duncan, 2014).

4 The CFS as a political battlefield

This chapter explores some of the unfolding dynamics arising from rural social movements' engagement in the CFS through the CSM, as they seek to participate in and simultaneously shape and contest this innovative arena, despite the many challenges. The chapter mainly presents examples from participant observation conducted during the negotiations of the CFS Principles on Responsible Agricultural Investments and Food Systems (CFS-rai), and the preceding preparatory meetings, in 2014.[1] The first opens with a presentation of the ongoing dispute and competing views over the meaning of 'responsible investments'.

It shows how food sovereignty, as a political framework, is utilised by members of the CSM group to negotiate and guide their political actions. It further demonstrates how LVC members strategically seek to use their engagement in the CFS to shape and politicize the policy processes. Based on the evidence presented in this chapter, I argue that LVC's engagement in the CFS goes beyond just winning support for advancing policy proposals. In fact, it is part of a broader struggle over identity, meaning-making and defining which types of knowledge and practices count as evidence and thus provide the basis for global policy-making.

These observations compel us to nuance the debate about what 'successful engagement' of social movements' participation in international institutions means. The chapter ends with a discussion on how strategic engagement plays out from within the CSM space and how participation of those most affected by food insecurity in the CFS injects new life into the debate about legitimacy, representativeness and the democratisation of transnational policy-making.

1 A world without – or investing in – peasant agriculture?

After almost four years of deliberations, preparation of drafts and documents, CSO internal negotiations, Skype meetings, strategising and lobby work, 300 stakeholders gathered in the CFS for the opening and first round of negotiations on 'The Principles for Responsible Investment in Agriculture and Food Systems' (CFS-rai). The FAO's green room hosting the reformed CFS was filled with not only UN country representatives, but also with spokespersons of small-scale farmers, agricultural workers, indigenous people as well as other representatives of sectors of society that historically have not had access to global policy-making processes. Here, social movement delegates sat side by side with UN ambassadors, private sector representatives and experts from international institutions, gathered in the FAO meeting room with earphones and microphones in front of them. All were ready to engage in international negotiations over investments in agriculture and food systems.

Although this first round of the CFS-rai negotiations took place between 19–24 May 2014, most of the delegates arriving in Rome, Italy for this week of negotiations had already been involved in this process[2] and previous CFS processes. To understand what was at stake in these negotiations, we need to go back a few years to the food and financial crisis that triggered a new wave of investments.

a The global land rush

As presented in the introduction of this book, the soaring food and fuel prices in 2007–2008 pushing more than 100 million people into poverty (World Bank, 2013) and leading to riots in over 60 countries forced world leaders to react. The demand for action was reinforced by the mounting concerns over the documented experiences of dispossession, violence, and social exclusion related to large-scale 'land acquisitions' and led a variety of experts, international institutions and policy makers to engage in debate on how to make the global food system fit for these challenges.

For some, large-scale foreign investment in developing countries' agriculture was presented as a means of stimulating food production and solving the food crisis (McKeon, 2015: 167). Consequently, the World Bank launched seven principles in January 2010 (see Appendix 3) that constituted the 'Principles for Responsible Agricultural Investment that Respects Rights, Livelihoods and Resources' (RAI) (later the 'PRAI'), jointly formulated by the World Bank, the International Fund for Agricultural Development (IFAD), the UN Conference on Trade and Development (UNCTAD) and the UN Food and Agriculture Organization (FAO), and endorsed and promoted by the group of G8 (McKeon, 2015: 167).

However, outside the meeting rooms of governments and international institutions, LVC and its allies in the broader food sovereignty movement did not accept that the principles for investments in agriculture should be formulated exclusively by a global elite behind closed doors (Gaarde and Hoegh-Jeppesen, 2011: 50). The approach to this new 'global land rush' (Borras and Franco, 2009; Margulis et al., 2013) was associated with investors and corporations intensifying their acquisitions and global competition for land, with very low levels of transparency, consultation, and respect for the rights of local communities living off the land (Borras and Franco, 2009; Cotula, 2011). In a public briefing released in April 2010, an alliance of 130 civil organisations,[3] including LVC, strongly rejected the seven principles as a way to legitimise the corporate (foreign and domestic) takeover of rural people's farmlands (FIAN, 2011). In February 2011, LVC and its allies of rural networks played a key role[4] in filling the old classroom at the Dakar World Social Forum (WSF) in 2011 to finalise the 'Dakar Appeal Against Land Grabbing' calling for an immediate stop to land grabbing (Field observation, World Social Forum Dakar, Senegal 25.02.203; see also LVC, 2011). This network drew attention to the need for small-scale farmers to be supported so as to feed themselves and their communities and to fight against these 'land grabs' (Borras and Franco, 2009).

In fact, the main dispute is between two competing views on and approach to investments. Briefly summarised, one perspective argues that foreign direct investment in agriculture and export of food crops is crucial for strengthening the agricultural sector and food security, which is why these should be promoted under regulated circumstances (states controlling or 'taming' irresponsible investments). The other perspective argues that large-scale investments in arable land are not the solution to global hunger. Instead support should be given to farmers, already investors themselves (economic, social, ecological investments). From this latter perspective, the support of small-scale farmers' investments is more beneficial to food production and local economic development, while the role of the state should be to guarantee that these conditions flourish.

The critique of the uncritical approach to large-scale investments did not only come from civil society and activist scholars but also from a number of international experts. This included the United Nations Special Rapporteur on the Right to Food, Olivier De Schutter, who strove to influence the discussion on land and investments from a human rights perspective. The analysis of De Schutter was that buying farmland and outsourcing food production has become a new strategy for governments that are worried about their future ability to feed their own countries, in order to achieve food security (De Schutter, 2011: 251). De Schutter raised particular concern over the development of large-scale, mechanised forms of agriculture, which tend to employ fewer farmers than traditional agricultural methods and are unlikely to support rural development and poverty alleviation: 'Considering the terms of the competition between large-scale and small-scale farms, in the absence of perfectly segmented markets, the risk is that the small farms will lose out' (De Schutter, 2011: 263).

Instead of promoting the PRAI, De Schutter recommended the development of a set of 'Human Rights Principles to Discipline Land Grabbing', stating that land-related investments should respect a minimum of 11 human rights principles to ensure that states are not tempted to consider large-scale agricultural investments from a purely economic standpoint (De Schutter, 2011: 263). In 2010 the reformed Committee resisted endorsing the PRAI (Duncan, 2014: 173)[5] and chose instead to begin an inclusive process with full participation of civil society and declared that it would 'help ensure that agricultural investment promotes food security' (CFS, 2010: para. 26). The following section presents how the two different views of investments presented above consist of a competing set of meanings.

b Contestation and negotiations over meanings

Entering the green room of the FAO building to observe the first round of the CFS-rai negotiations in May 2014, the seriousness of the opening of the negotiations permeated the room. I took a seat in one of the rows dedicated to the CSM and I sat between a younger indigenous woman and an experienced LVC leader from Asia. As the LVC member seated next to me summed it up, 'There is an awful lot at play here.' The Peasant leader from South Asia elaborated this with the following account.

> I am a farmer but agriculture has been turned into a losing way of life: our lands, the main resource for our livelihoods, are being grabbed for private corporations (…) We have been trying to get the attention of the government but they have no time for the farmers, they make false verbal commitment. The governments cannot grab our land for private corporations at any cost (…) In my country the government can acquire the lands of farmers for the benefit of industries. The land can be distributed to the industries. Small and marginal farmers are selling out and quitting farming. Also, even big farmers keep selling their farms because of the crisis in agriculture. Big investors convince people that it is better to make money; it is better than doing farming. We don't need investments in land that drives farmers away from their livelihoods. We need investments in people living on the land. We are here to fight but for our right to farm and feed our communities with healthy food.
>
> (South Asian member of LVC, CFS-rai negotiations, Rome 25.05.2015)

Additionally, an Indigenous Peoples' spokesperson – engaged in global CFS negotiations for the first time – reiterated how investments have a direct impact on the livelihoods of many rural people around the world.

> Many governments are supporting large-scale investments in land and present them as development opportunities. But land is being turned into an attractive commodity for investors. These types of investments have led to displacement, insecurity, and environmental degradation. It is our plight here to reveal what this really means and what is actually happening on the ground.
>
> (Indigenous Peoples' spokesperson to the CSM, conversation, 20.05.2014)

The importance of clarifying the very meaning of investment was also emphasised by a West African peasant leader during an interview at a workshop on land grabbing at the World Social Forum in Dakar, Senegal. The LVC leader explained: 'We need investments. But let us be clear on what we are talking about. The concept of investment needs to be reclaimed!' (African LVC member interview, World Social Forum, Dakar, Senegal, 15.02.2011). In a similar vein of thought, another LVC member emphasised, during his participation in the CFS-rai negotiations in 2014, that the practices of industrial farming and investments are not 'neutral' but have direct consequences for people of the land.

> There is a widespread mythology that investments are good and the more investment the better. But investments are not innocent. The problem is that investors convince us to sell our land for textile industries. The problem is if one farmer sets up a dyeing industry that pollutes the waters, the land of the neighbouring farms may also become useless.
>
> (LVC member in the CSM preparatory meeting for the CFS-rai negotiation, 15.05.2014)

Throughout the CFS-rai negotiations, the civil society group raised questions such as: Who will benefit from these investments? What do investments mean for social objectives? LVC's struggle over investment is an example of a discursive struggle and an attempt to 'fill in meanings' into concepts in correspondence with their own specific values (Laclau and Mouffe, 2001: 122). When social movement activists present their own interpretations of notions such as 'responsible investments', 'efficiency', and 'sustainable', and contest messages that are delivered as 'objectivity' and 'neutrality', they actively challenge the dominant discourse that is often presented as common sense (McMichael, 2010: 3).

The presentation of an alternative analysis of investments by LVC and its allies is an example of the social questioning of the dominant 'episteme' 'based on the top-down market calculus where the market has become the dominant lens through which development is viewed' (McMichael, 2010: 3). For instance, during the CFS-rai negotiations, LVC strongly contended that investments are not just about the mobilisation of financial resources, but also about the commitment of multiple resources (financial, natural, human, social, cultural) to realise a range of goals that do not simply focus on economic return. During a round of the CFS-rai investments, a South-Asian LVC member clarified the importance of mobilising strongly against the logic of the market based on 'quick-wins' and 'market fixes'.[6]

> Investments are not only for economic benefits. Investments may have a social and ecological function for local communities. For instance, investments can mean taking care of the soil, the welfare in your community or the environmental system. In the agroecology model, cash is only a small part of the agroecological economy (…) when we invest in agroecology, we invest in a way of life. It is about investment that fosters values, solidarity and care for the environment and the well-being of your neighbours and the next generations to come. This contrasts the market calculus where every centimetre is cultivated to maximise profit.
>
> (European LVC member, interview, CFS41 Annual Session, Rome, 10.10.2014)

During the CFS-rai negotiations the members of the civil society mechanism emphasised that *all* investments should be responsible. During negotiations CSM members presented the argument that investments are only responsible if they prioritise the rights of small-scale producers and workers and uphold the rights of small-scale food producers, indigenous peoples, and local communities to access, use, and have control over land, water, and other natural resources (Observations, CFS-rai negotiations, 19–25 May 2014). The list of the key positions taken on the CSM on investment in agriculture is presented in Box 4.1.

Box 4.1 CSM policies on investment in agriculture

1. Investments must contribute to and be consistent with the progressive realization of the right to adequate and nutritious food for all.
2. Investments in food and agriculture must ensure protection of ecosystems and environments.
3. All investments in food and agriculture must ensure decent jobs, respect workers' rights, and adhere to core labor standards and obligations as defined by the International Labour Organization (ILO).
4. All investments in agriculture and food systems must ensure decent incomes, livelihoods, and equitable development opportunities for local communities, especially for rural youth, women, and indigenous peoples.
5. Investments must respect and uphold the rights of small-scale food producers, indigenous peoples, and local communities to access, use, and have control over land, water, and other natural resources.
6. All investments must respect the rights of indigenous peoples to their territories and ancestral domains, cultural heritage and landscapes, and traditional knowledge and practices.
7. All investments must respect women's rights and prioritize women in benefit sharing.
8. States must mobilize public investments and public policies in support of small-scale food producers and workers. Small-scale food producers, workers, and their organizations must be meaningfully involved in the formulation, implementation, monitoring, and review of these investments and policies.
9. States must protect small-scale producers and workers from market fluctuations and price volatility by regulating local, national, regional, and international food markets and curbing food price speculation.
10. States must respect and support timely and nondiscriminatory access by small-scale producers, workers, indigenous communities, local communities, and the public to justice; grievance mechanisms; fair, effective, and timely mediation; administrative and judicial remedies; and a right to appeal.
11. Trade and investment agreements and treaties must not undermine or compromise the rights of small-scale food producers, workers,

indigenous peoples, or food sovereignty. States must monitor and assess the impacts of such agreements on the realization of the right to food and take appropriate action where necessary including through renegotiation or cancellation of the agreements/treaties.

12 States should enact appropriate national laws to regulate and monitor extra-territorial investments and investors. In so doing, they should apply the Maastricht Principles on Extraterritorial Obligations of States in the Area of Economic, Social, and Cultural Rights, as the guiding document.

13 The effective, meaningful, and democratic participation of small-scale food producers, workers, and indigenous peoples, particularly women, must be guaranteed in the planning and decision making around agricultural investments, area development, and land and resource use and management.(CSM, 2014b)

The analysis above, coupled with participant observation and interviews conducted within the CSM (2012–2014), have led to the identification of food sovereignty as the central political frame when civil society activists come together in the CSM to develop joint strategies and positions in the context of the CFS. This is the focus of the following section.

c Building policies for food sovereignty

Although views on food sovereignty differ among the actors within the CSM group, the principles of food sovereignty are increasingly shared by members in the CSM group and provide an overarching framework for action that allows for incorporation of various issues (McKeon, 2015: 115; Duncan, 2014: 160). As a political framework, food sovereignty is largely developed to politicise food security and other aspects of agriculture (McMichael, 2014; McKeon, 2015; Brem-Wilson, 2015; Desmarais, 2007).[7]

Even if the term food sovereignty is not incorporated into CFS outcome documents, some of the food sovereignty principles have been included as a result of successful civil society lobbying (Duncan, 2015: 160). For instance, during the intensive CFS-rai negotiations, LVC and its allies fiercely defended the argument that small-scale producers should be – and already are – the main investors in agriculture.[8] Supporting smallholders to invest in their own agriculture is a key component of building food sovereignty (McMichael and Müller, 2014). Field observations confirm that CSM members actively use food sovereignty to politicise the policy debate and to address the rights to be realised, as well as the roles and responsibilities of different actors in the food system (see also Brem-Wilson, 2015: 10; McKeon, 2015).

One way social movements are seeking to politicise the debate is by raising concerns related to issues of distributional justice, fairness and access to decisions and resources. These can be observed when members of the CSM critically ask questions such as: Who has access to land? How is land used? And, for which purpose? (Observations, CSM, preparations of the CFS-rai negotiations, May 2014).

Members of the CSM group often reiterate that investments are already being done by farmers themselves (economic, social, ecological investments). From this perspective supporting investments of small-scale farmers themselves is more beneficial to food production and local economic development, while the role of the state should guarantee the conditions for these to flourish (cf. Box 4.1; see also McKeon, 2015; McMichael and Müller, 2014).

While small producers are economic actors themselves ('the main investors in agriculture', HLPE, 2013), peasants delegated to the CSM often seek to politicise the debate by emphasising the distinct practices such as agroecology, solidarity, care for the environment, public interests, the well-being of the next generations and contributions to eradicate hunger. From this viewpoint members of the CSM group argue that the priority logically should be given to strengthening and securing smallholders' own investments and they reiterate that it is the responsibility of states to support the empowerment of small-scale producers and workers to strengthen their knowledge and skills (Observations, CFS-rai negotiations, 2012–2014; see also McMichael and Müller, 2014).

d Defining roles and responsibilities

Observations of the ongoing dynamics in the CFS show that when activists engage in the policy arena they actively seek to unpack the discourse of national governments, which are often considered by social rural movements to be the 'duty bearers' and therefore responsible for failed land policies (Borras and Franco, 2009: 28).

> Everyday people around the world are suffering because governments around the world undermine national food security. Hunger is not about scarcity; it is about injustice. The responsibility and priority of states to feed the people cannot be negotiated.
>
> (European LVC member, CFS-rai negotiations 10.08.2014)

LVC members are not afraid of criticising governments if they find that elected policy-makers are failing to fulfil their mission of defending common interests and public goods. As one South-Asian LVC member put it in an interview during a CFS session:

> We all know that the national short-term goals and economic interests of governments conflict with the long-term goal of combating global hunger. Governments work for re-election and there is more money to get from the private sector.
>
> (South-Asian LVC member, interview, CFS Annual Session, Rome. 15.10.2014)

Adams and Tobin have argued that government representatives often 'rather than representing the breadth of commitments, interests, and dialogue of their countries, they are trying to minimise their commitments and maximise their short-term gains' (2015: 23–24). One representative from an International NGO following the CFS reform process closely, in an interview after a CFS roundtable debate suggested that the members of the CSM are embracing a 'citizen's perspective' and function as a 'moral compass' in the CFS arena (CSM observer, personal conversation, CFS Annual Session 15.10.2013). Civil society actors often remind governments about their obligations to combat global hunger, fulfil human rights[9] with strong appeals to values of 'moral intuitions, felt obligations and rights' (Goodwin et al., 2009: 13).

2 Mobilising institutions – institutionalising movements[10]

a Political manoeuvring

When mobilised peasants engage in the CFS they strive to strategically use their engagement to shape the processes and the arena in which they engage. They seek to define the terms of the debate and the power balance in the arena to their own advantage. De Certeau defines 'strategy' as 'the calculation of power relations that becomes possible when subjects with a will of their own act from a vantage point that is their own and that serves as a basis for their relations with a separate and distinct exteriority' (de Certeau, 1990: xlvi). In contrast, when a person cannot count on a space of his or her own, or cannot maintain a clear boundary to distinguish himself or herself from the others as a clearly visible totality, then that person can only make 'tactical calculations' (ibid.).

Data from fieldwork observations within the CSM show that the CSM is clearly a strategic space where power analyses are made, where both strategy and tactics are discussed and decided. In the CSM space, civil society actors engaging in the CFS seek to develop their own – to use Tarrow's (1998) term – 'action repertoires', ranging from debate to confrontation. Dialogues, lobbying and advocacy work to more confrontational methods, such as 'walk out' protests, 'naming and shaming'[11] are some of the tactics discussed by members (Observation, CSM preparatory meeting, Rome, ahead of the 40th Annual Session of the CFS).

During CSM meetings, social movements delegated to the CFS often reiterate that their engagement in the CFS cannot merely be tactical, but has to become *strategic* to shape the directions of the negotiations. However, while members of the CSM group employ a lot of time and energy in their autonomous civil society space trying to pre-empt and re-orient the agenda, there is constant danger that the civil society group will end up acting in a predominantly tactical manner by 'running after the agenda set by other stakeholders':

> The most interesting, promising and mobilising advances have emerged when social movements are proactive. We need to look at things from our own perspectives rather than within the frameworks set by governments.

> We cannot only do tactics. We need to think about the big picture here. This is very strategic. We have created a political space for putting our own agenda on the table.
>
> (European LVC member, CSM meeting on 'food sovereignty, seeds and agroecology', CSM Annual Forum, 5.10.2014)

This was also one of the clear messages conveyed by the self-evaluation (2012) that the LVC made of its first two years of engagement in the CFS: social movements must be strategic and proactive rather than defensive. This assessment clearly spelled out the challenges for social movements wishing to engage in a UN arena with inherent power imbalances: 'If we accept the logic of negotiation, we must for example be ready to surprise governments by manoeuvring them into areas where they do not expect to find us in order to rearrange the balance of power in our favour' (LVC, 2012: 12).

Members in the CSM carefully calculate their technical and political interventions in the CFS arena as part of an over-all strategy. Technical interventions are often statements built on well-prepared 'counter-facts' presented as strong political intervention addressing governments' roles and responsibilities. Technical support persons in the CSM groups may at times support the technical formulation of these statements. The support provided by this technical capacity contributes to the strategic engagement within the group, as these persons are often familiar with UN texts and know how to develop priority areas so as to defend key issues of the CSM.

The battle over political manoeuvring is registered when some stakeholders in the CFS arena start to move the debate into a question about technicalities.[12] Social movement delegates are often those who do the political interventions on behalf of the CSM group, by presenting strong messages with appeals to government responsibilities. This was, for instance, observed during a CFS-rai session where a LVC member took the microphone to address the responsibility of governments for their actions or non-actions:

> Governments should not get away with turning issues of access to land and natural resources into technical issues. These are political questions. For those who endure daily struggles against injustice, oppression, poverty, displacement, access to land and livelihood is a question of life and dead. If there are not actions and people continue to go hungry with these policies, this is a government responsibility.
>
> (European LVC member, CFS Session Rome, 16.05.2013)

LVC members engaged in the CFS often emphasize that they engage fully in the policy process until governments take the final decisions. As was explained in Chapter 3, government responsibility was a key reason why LVC and its allies during the CFS reform process strongly advocated for civil society to have a voice but not a vote, so as to be able to hold governments responsible as the final decision makers.

b The constant 'back and forth'

Despite the awareness of the need for strategic action, engaging in intergovernmental negotiations in the UN arena is a constant 'back and forth[13] for social movements. For instance, while social movements are capable of acting strategically and constantly seek to influence the agenda, sometimes they are also obliged to act tactically. Actors 'constantly need to play with events in order to transform them into opportunities for making an impact that may not last' (Müller, 2011: 248). Other constraints of the UN system that may oblige civil society actors to turn to these tactics are exemplified in the fact that negotiations often turn into a process of 'damage control', e.g. to defend results already achieved in earlier negotiations. Such 'reactive repertoires' – to use Tilly's term (1978: 144)[14] – can be draining for activists who invest energy in building alternatives and 'raising the bar' for human rights standards in the CFS. An NGO representative reported this in an interview during a lunch break at the CFS-rai negotiations:

> Negotiations can sometimes feel like being in a long tunnel, where it can be difficult to see the light at the end. This time, I felt that our role as civil society during negotiations was very much about damage control. For instance, when we went into the endless debate on the FPIC[15] again, I felt we were rolling back 30 years of struggle for indigenous people's rights. Are we really taking steps forward or backwards here?
>
> (NGO representative, field notes, CFS-rai negotiations, 10.08.2014)

Although the UN institutional environment exerts an influence on the action repertoire of civil society and social movements' organisations, this does not mean that engagement is reduced to purely defensive tactics. As explained above, engagement is a constant struggle over political manoeuvring. In this regard, data from observing social movement activists that are engaged in the CFS show that the CSM space helps members to step out of the UN logic and move into its own space to strategise behind closed doors for members of the CSM group (and those that occasionally are invited to participate in this space).[16]

Clearly, members of the CSM group seek to undertake strategic action at the CFS. A case in point is their advanced lobbying and collective analysis of the positions of governments during CFS negotiations. One NGO member of the CSM explained the importance of carefully following the political interventions in the CFS so as to 'reveal the contradictions between governments' statements and policies on the ground' (CSM member, interview, CFS Annual Forum, 10.10.2013). Such observations disclose that civil society actors are becoming increasingly savvy in the processes and procedures of the Committee and are, as a result, becoming more influential and effective in negotiations (Duncan, 2014).[17] Yet, despite these efforts to build a strategic space from which civil society actors can autonomously self-organise their actions, when they move into the CFS they become exposed to the 'classical risks' associated with institutionalisation, such as co-optation and de-politicisation.

c *Politicisation versus de-politicisation*

Several scholars have warned that when social movement activists get access to formal political institutions this inclusion will be at the expense of more radical policy changes (Gamson, 1990; Meyer and Tarrow, 1998; Smith and Weist, 2012: 171). The observations presented in this book challenge this assumption, although they do not negate the risks of 'taming' (Kaldor, 2003) and co-optation that social movements face as they engage with institutions. Rather, they illustrate the ongoing dynamics between institutions and social actors. This relationship between social movements and institutions as a dialectic process is well-captured in the title of the study 'Mobilising institutions – institutionalising movements' (Müller and Neveu, 2002), which explores these new forms of relationships between the state, institutions and civil society participation.

Scholars of anthropology remind us that engaging in the realm of institutions often entails subtle processes: 'Actors are simultaneously instituting' and being 'instituted', which is to say that 'they participate in the production of norms and rules and are themselves integrated into the system of constraints' (Müller, 2011: 183).

In the context of exploring social movements' participation in the CFS, this ongoing subtle power struggle was expressed by a Latin American LVC member during a CFS session:

> On the one hand, with our participation, we are influencing the processes, but we must be alert: When is our participation starting to sustain the governance of industrialised agriculture in the world? And when do we impact them and when do they impact us?
>
> (Latin American member of LVC, interview, CFS-rai negotiations, Rome, 17.05.2014)

This subtle tension between politicisation and de-politicisation is particularly noticeable in the ongoing struggle over agenda setting within the policy arena. For instance, LVC presents central terms such as food sovereignty and agroecology as a vision directed towards a radical transformation of society. By bringing 'radical' issues to the table and seeking to politicise the debate in the CFS, there is an inherent risk of both co-optation and de-politicisation. This level of consciousness is conveyed when LVC leaders engage in the CFS arena. For instance, one LVC member representing the agroecology movement in Latin American stated his fear that other stakeholders were starting to use self-interpreted 'market interests'.

> We must protect our terms from market interpretation. We are aware that agroecology may be the new tool in the toolbox of industrial agriculture.

> We must be vigilant of agroecology not being green-washed and used to legitimise industrial agriculture.
>
> (Latin American member of LVC, Skype interview, 30.10.2014)

Another LVC member echoed this concern when I interviewed him after the *International Symposium on Agroecology for Food and Nutritional Security*,[18] held at the FAO in September 2014:

> Even though governments start to open up, we must remain watchful here and at home. It is up to us to hold them accountable for the commitments and monitor that this is not being used to legitimise other forms of agriculture on the ground.
>
> (Latin American LVC member, interview, Rome, 30.09.2014)

This human consciousness underpins the overall attempt to retain a strategic calculus, suggesting that the challenges related to co-optation and de-politisation have been overcome at different times. As expressed by one LVC staff support person:

> The risk of co-optation is always there. For instance, if we move to debate agroecology in the CFS, and if we are not very active and well organised, the governments may then pick up our concept, turn it around and fit it into their own concepts and use it to support their own practices.
>
> (Skype interview, 30.05.2014)

As this LVC staff support person suggests, members (and supporters of) the movement will need to be active and offensive in the debate so as not to fall into the trap of legitimising other practices, which serve as tools that potentially sustain the status quo of the dominant industrial agricultural model. In their engagement in UN processes, social movements face the constant risk that their participation may end up legitimising policies that they do not necessarily agree with.

Whereas it is an ongoing struggle to remain active in the debates and retain the strategic lead and to politicise the discussions in the CFS arena, LVC's engagement in the CFS demonstrates that social movements are not automatically becoming the 'pawn of the processes' (Duncan, 2014: 149) when they engage in a formal UN arena. Social movement leaders are increasingly becoming more skilled in manoeuvring within the system, where engagement is complex and dynamic, and not automatically negative. The next section further debates the meaning of social movements' engagement in a global policy and discusses what 'success' may mean in this context.

3 Social movements' successes beyond political influence

The last part of this chapter presents how the presence of direct representatives of peoples' organisations in the UN arena brings new debates related to legitimacy and representativeness. It also discusses what 'successful engagement' means for a rural social movement engaging in a global policy arena.

a The peasant identify as a site of struggle

When social movements move into the CFS arena, a central component of contestation involves challenging the socially prescribed identities. The articulation of a peasant identity across borders is a strong political act for peasant members of LVC (Desmarais, 2007: 197). 'We really like to define ourselves here' is a phrase that is often repeated by LVC members during CFS sessions (Field notes, CSM preparations ahead of the CFS-rai negotiations, Rome, 23.04.2013). According to Alvarez, Dagnino and Escobar, the notion of identity and forms of subjectivity that we inhabit play a crucial part in determining whether we accept or contest existing power relations: 'For marginalised and oppressed groups, the construction of a new and resistant identity is a key dimension of a wider political struggle to transform society' (1998: 5–6). In the context of the CFS, resistance can emerge when social actors intervene in intergovernmental processes, question social norms and appropriate the language in order to shift the discourse of 'victims' or 'poor farmers' towards an understanding of peasants as key agents of social change and producers of society.

> We do not want to be saved by industrial agricultural practices invented by farmers in the North. What we need is support for peasant-based agriculture. This is the model that not only peasants, but humanity, depend on. We want to be recognised for what we are: main investors in and producers of agriculture.
>
> (Latin American LVC member, CFS40 Annual, Rome 13.10.2013)

Peasant delegates to the CFS interviewed for this study often emphasise that they belong to a peasant category that is separate from the market orientation of agriculture (McMichael, 2010). Yet, One Latin American member reported in an interview that governments often seek to dilute differences and conflicts in their countries when they talk about farmers as one category. During a CFS plenary session the LVC member openly criticised the other actors in the CFS for the 'obscured deliberate fuzzy categorising of farmers':

> We are not farmers in harmonious villages. Farmers engage in bloody struggles around the world. Governments know that, but they often close their eyes, as they need to support proposals for the agro-industry. We don't want to be lumped together with agribusiness farmers. We represent the work of honest people, not companies and private interests.
>
> (Latin-American LVC member, CFS-rai negotiations, Rome, 18.05.2014)

As the previous chapter demonstrated, the CFS declares itself to be a 'multi-stakeholder platform that enables all viewpoints to be considered' (CFS, 2011) and gives the same rights to all stakeholders (states, civil society organisations and the private sector) in the processes leading to global policy-making. However, observations conducted in the CFS arena reveal that the term 'multi-stakeholder platform' remains largely contested by social movements and other civil society actors. From the viewpoint of many peasant activists, the CFS term 'stakeholder' alludes to the idea that all stakeholders have the same weight and thus dilutes power imbalances and different mandates (McKeon, 2015). A European LVC member expressed how this was highly problematic during a CSM session at the Annual Forum in 2014: 'We are not equal stakeholders with the same roles and responsibilities. The term stakeholder is a term used in the business world. Stakeholderism is a name that dilutes power struggles and is not a term that we use' (European LVC member, CSM Annual Forum, Rome, 10.10.2015).

Members of the CSM group often argue that instead of striving towards 'multi-stakeholder governance' the CFS should rather aim for 'multi-actor governance' (Field observations; CSM Annual Session, 10.10.2015). Zanella and Duncan summarise the latter as

> taking a rights-based approach, where states would then be able to play a decisive role in balancing the power asymmetries that do exist amongst all actors, in particular between large corporations and small and medium size enterprises, between smallholder and larger farmers.
>
> (Zanella and Duncan, 2015: 3)

Nevertheless, representatives of corporations and agribusiness farmers are increasingly moving into the CFS seeking to maintain the dominance of global food and nutrition policies (McKeon, 2015; Zanella and Duncan, 2015). The constant tension between peasant farmers and agribusiness farmers reveals a deeper power-struggle within the policy arena, especially in the CFS, where we see multiple farming actors fighting to gain legitimacy. As expressed by an African LVC member: 'We do not want the science from Western farmers. We do not need that kind of technology and knowledge. We need support to develop our own knowledge and practices' (African LVC member, interview, CFS Annual Session, Rome, 15.10.2013).

b The struggle over knowledge and the right to produce society

When peasant delegates engage in the CFS they do not only seek to change the policy outcomes of the UN processes; they also engage in a struggle to challenge terms of the debate and politicise the debate in a larger struggle over meanings, and who has the right to be included in defining these meanings. This is an example of social movements engaging in what Alvarez, Dagnino and Escobar (1998) call 'cultural politics' that relates to who has the power to fundamentally determine the meanings of social practices. As Desmarais states:

'The Vía Campesina's cultural politics redefines what it means to be a peasant or a farmer, redefines what constitutes knowledge and who gets to define and control knowledge, introduces new concepts, and hence helps to shape the international agenda' (2007: 237).

A number of scholars analysing LVC and other transnational agrarian movements have documented the struggle of peasants, who have pursued an ontological alternative to the dominant model of 'western modernisation' by seeking new ways of 'being and existence' (McMichael, 2006: 220; Edelman, 1999: 102–103; Desmarais, 2007: 197).[19] For instance, McMichael argues that when social agrarian movements are struggling to assert their identity they claim an alternative ontology to the market ontology of the capitalist modernity, where peasants are no longer being considered producers of knowledge (McMichael, 2008: 3; see also Desmarais, 2007: McKeon, 2015).

The struggle for peasants to gain recognition for their knowledge and practices has often been in direct opposition to Eurocentric frames of 'northern epistemologies' (Santos, 2007). In this regard the rise of peasant activism can be seen as an example of a former marginalised group of society contesting the 'universalising knowledge dominated by the hegemonic imaginary of the North'[20] (Escobar, 2001: 149). This is the backbone of social movements claiming to validate people's own epistemologies (Santos, 2007). As expressed in an edition of the international Nyéléni Newsletter,[21] it is not only central for peasants, but also for indigenous peoples, pastoralists and other historically excluded actors of society to present their own interpretations as legitimate forms of knowledge:

> The reality is that there is a real attack underway against our territorial memory, our memory of place – the lands which are our vital surroundings, our common environment; we need to recreate and transform our existence: the spaces we give meaning to with our shared wisdom and knowledge, with our common history.
>
> (Nyéléni Newsletter, 2014)

For LVC, challenging the dominant worldviews on what certain terms mean, such as expertise, development and progress, is a cornerstone of food sovereignty. The attempt to win recognition for locally based knowledge and the rights of peoples and communities to define their own food systems is thus part of the larger peasant struggle against the dominant agribusiness model. As one African LVC peasant leader phrased it during the 40th Annual Meeting of the CFS:

> Our voices cannot longer be ignored. The converging crises have clearly documented how the dominant model and intensified grabbing of peasants' land, seeds and resources undermine the diversity, the knowledge and practices of those producing most food for society.
>
> (African LVC member, CFS, Rome, 12.10.2014)

The struggle over which forms of production and practices count as 'evidence' and thus as a foundation for global policy-making on food-related issues remains an ongoing struggle (McKeon, 2009a: 188). In this regard the High Level Panel of Experts, established as part of the CFS reform requires special attention (HLPE, 2013).[22] At the global scale, the HLPE is a unique scientific body that respects the plurality of knowledge systems, and has proven to actively extend the concept of research to include non-academic sources, such as knowledge from social movements. The HLPE can be seen as a new constellation for knowledge production that increasingly gives visibility to the different forms of knowledge and practices that emerge in the world's peripheries (McKeon, 2015: 190; HLPE, 2013). We return to the role of the HLPE as a scientific platform and the issue of co-building of knowledge in the concluding section of this book.

c The success of social movements is not mathematical

While a key aspiration of social movements' engagement in the CFS is to influence policy processes, their involvement in UN negotiations does not always mean that they leave UN negotiations with a strong set of negotiated principles that can be used to serve peoples' struggles on the ground. One LVC member, engaged in the negotiations, reported this in a conversation after the governments endorsed the CFS-rai document:

> In this processes we did not end with a strong set of investment principles for our members on the ground. I think these negotiations and the power imbalances that were revealed mirror very well the reality of our daily struggles on the ground. It is evident how manipulated some world governments are by big corporate interests. When you look at who normally dominates these spaces it just confirms that it is all the more important for small farmers to occupy this space to promote progressive policies that can help to build the food sovereignty vision. It emphasises the importance of being here. We will continue to fight at all levels in all spaces. It is not easy but we have no other choice than to continue doing what we are doing. We need to continue to mobilise here and elsewhere.
>
> (European LVC member, Rome, CFS, 14.05.2014)

One LVC member stated in his intervention during the CSM Annual Forum in 2014 that the premise of engaging in UN negotiations is that 'sometimes you lose, and sometimes you win':

> We must face the reality that sometimes the outcome here is not always directly useful on the ground. Sometimes we lose a process, but we may also gain a lot. We learn to reflect on how we work together and progress. We know how to be better organised and what their tactics are. We get to know both our enemies and ourselves better from this engagement. We

> get stronger when we learn to work better together. We show that other stakeholders cannot divide us here. If they divide us, then we lose. We must remind ourselves that our efforts here are only small but important pieces of the larger struggle to build the model we wish to see in the world. We are creating something new here. Of course we are stumbling on the way. We have major obstacles ahead of us, but we are getting stronger in each encounter.
>
> (European, LVC member, CFS-rai negotiations, Rome, 17.08.2014)

Higgott (2001: 135) has emphasised that we need to be careful not to judge the engagement of social actors in the UN from predefined efficiency and effectiveness categories. 'Effectiveness' is often too narrowly defined in terms of whether or not the engagement will influence key decision makers. As McKeon asserts, the success of social movements is not 'mathematical':

> The main aim of social organisations is not to have an X and Y UN outcome. Rather, it is to use the space of global governance that the UN potentially affords to help locally based movements build upward and outward links that can amplify their voices and collectively project alternative proposals to the liberalisation agenda.
>
> (McKeon, 2009a: 176)

Beyond seeking to influence the actual policy processes at stake in the CFS, one can imagine different aspects and degrees of success for a social movement engaging in an intergovernmental institution. Drawn from the observations and interviews with civil society participants engaging the CFS arena, I provide some examples. For instance, this book has demonstrated how social movements, moving into intergovernmental institutions to engage in deliberations and negations, build new relations and networks. From these interactions and development we could think of *collective network success*. Kinder (1998) argues that effectiveness may be judged not on the basis of perceived influence per se, but rather in terms of whether a movement is successful in strengthening solidarity and strategic connections within the group. Therefore, Kinder argues, instead of focusing on the short-term goals, people can be motivated by the perceived success of the rally in consolidating the movement, with a view on implementing change in the medium and long term which may be obscured by narrow definitions of effectiveness (Kinder, 1998).

As was emphasized in the very beginning of this chapter, building a common frame for action and finding ways to work together in diversity is also important to ensure the success of strong global food movement outside the halls of the UN.

Another example of success is the appropriation of new instruments and techniques (lobby and advocacy, monitoring instruments) that are developed from strengthened participation in the transnational arena. In this regard, the ongoing manoeuvring techniques that are developed through the discursive struggle in the policy arena may help social movements to succeed in shifting the debate within the arena. Social movements seeking agenda control and policies that can lead to positive changes on the ground. If the CFS arena can help to mobilise government support for social movements proposals, this is positive step. While social movements proposals do not always gain traction in the intergovernmental UN arena, civil society actors may use the CFS arena to build support from governments in other areas. Taking into consideration that some governments are not open to dialogue with civil society in domestic arenas it can be crucial for CSO representatives to use the direct interaction with governments in CFS to open up dialogues and processes for people to participate in policy processes 'back home'. As a member put it in an interview during a CFS session:

> We are learning to build the links here. We need to create mechanisms to check which governments violate the rights of peoples around the world. Even though we only reach guiding principles, they can provide protection for communities, if we keep insisting on these things on the ground.
>
> (African LVC member, interview, CFS-rai negotiations, Rome, 05.08.2014)

Whereas UN documents such as 'guidelines' and 'principles' remain non-binding, civil society actors hope that engagement in CFS global policy processes can serve as a tool to exert pressure on national governments and hold them accountable in order to support the struggles 'back home'. The use of the UN arena as a potential global–local bridge-builder (McKeon, 2013: 105–122) is in line with Keck and Sikkink's 'boomerang model', which outlines how place-based movements and transnational networks and practices can pressure national governments to change their practices and align with international norms (1998: 13). Observations of the dynamics in the CFS show that civil society actors do not only lobby their national governments, they also lobby and build alliances with other governments. Activists seeking to exert pressure on a global scale in order to align with international human rights standards is an example of how civil society groups, sometimes blocked from making demands directly to the states 'at home', may look outside state borders to establish contacts and dialogue spaces with governments on agrarian policies, land conflicts etc.

Finally, engagement in a global policy arena can also lead to individual success when an actor increased his or her confidence and capacity to act in the policy arena. We return to the opportunities and challenges for people's agency to play out in the CFS arena, in Chapter 5.

4 Peoples' participation in global decision-making

This last section of this Chapter presents some of the debates arising from the direct engagement of representatives of rural constituencies – and other sectors most food insecure – in the policy processes of a UN Committee, hitherto restricted to governments.

a Legitimacy and representativity

Issues of legitimacy are increasingly addressed in the literature on civil society engagement in policy-making, notably around issues of the legitimacy of non-elected actors to participate in democratic processes (Buchanan and Keohane, 2006; Kersbergen and Waarden, 2004; McKeon, 2009, 2015; Duncan, 2015; Brem-Wilson, 2015). There is common understanding among stakeholders that while the CFS does not have formal power (decisions are non-binding), it has strong legitimacy power, in particular due to its nature as a multi-actor forum with a broad range of participants and because most decisions are achieved by consensus.[23] In the CFS the legitimacy of food producers comes from their ability to impact the governance processes from a personal perspective and experience (Duncan, 2014: 161). When peasants bring their stories from the field – derived from their personal experiences – to the CFS, they bring to the table issues of legitimacy as well as representativity:

> Governments do not see the reality from their office windows. Those with only desktop experiences that are remote from the everyday struggles on the ground do not know what it actually means to be affected directly by food insecurity. Peasants are those who live through the struggles of oppression, poverty or injustice or those close to those suffering from hunger and food insecurity.
>
> (CSM Spokesperson, representatives from the international workers movement, observations, CFS40, Annual session, 13.10.2014)

During an interview at the CFS-rai negotiations, a Latin American government representative emphasised that the CSM group has high legitimacy in the CFS arena: 'There are no arguments against those who have lived the struggle' (Latin American UN Diplomat, interview, CFS meeting, Rome, 17.08.2014). The relevance of farmer's lived experience and technical knowledge to the legitimacy of CFS is reflected in the question asked by the CFS-rai Chair during the opening session: 'How many farmers are in the room – can you please raise your hand?' (CFS Chair, Opening session, CFS40 Annual session, 2013).

A European UN diplomat stated in an interview that the CSM group plays an important role because it reminds governments of the role and mandate of the UN Committee.

> The strong testimonials from civil society actors brought in from the field remind us why we are here at the CFS. Such stories bring pictures to mind in these technical negotiations and make us remember why we are here: namely to combat hunger in the world and to ensure food security and

> nutrition for all. The civil society group is always well-prepared and it helps us to better strive for solutions that are favourable to all of us. My impression is that the CSM group is very admired by other stakeholders, in particular for their strong dedication and active contribution to the work of the CFS.
>
> (European UN diplomat, interview during the CFS40, Annual Session, 18.10.2013)

Government representatives to the CFS often acknowledge in speeches and statements that the authority and legitimacy of the CFS are strongly linked to the direct participation of those who are most affected by food insecurity. This means that small-scale food producers organized in LVC, a movement that counts approximately 200 million peasants farmers in 73 countries around the world, brings with it considerable influential legitimacy when their representatives enter an institution with the mandate to address issues of global food security. A food activist, member of the CSM and pending member to the LVC, explained that in his opinion the rural segment has a strong legitimacy power in the CFS:

> Even if government representatives do not admit it, they can be afraid of those representing peoples' organisations that are now in the room. They cannot just say what they want to. They know that we are watching them and represent significant segments of the population that governments may simply not reach.
>
> (Member of the CSM, personal conversation Rome, 13.10.2014)

Such ongoing power dynamics can be strongly felt in the room. For instance, by carefully observing how social actors engage with each other and react to each other's proposals during and between official sessions. This includes noting the body language of government representatives, whether it is the nodding of heads when peasant members of LVC intervene in the CFS plenary sessions, or the body language when UN diplomats approach LVC members between sessions in the policy arena.

In the context of LVC's engagement in the CFS, the transparent participation processes facilitated by the CSM reinforce the legitimacy of social movement actors (Duncan, 2014: 161). Field observations reveal that CSM members are remarkably well prepared when they engage in the CFS. As presented in Chapter 3, before and after CFS sessions, members of the CSM meet for 'preparatory' and 'wrap up' meetings. During CFS Plenary sessions members of the civil society group often bring with them piles of copies of their common deliberated and drafted CSM proposals that they give to governments as an attempt to gain their support (Field observations, CFS-rai negotiations, 20.05.2014).

In the breaks between CFS plenary sessions, civil society members organise their time efficiently to lobby and advocate for their positions. They either meet for internal strategising and evaluations or conduct different lobbying activities, such as lunch meetings with country delegates, regional group meetings or informal meetings where they seek to win support for their proposals. As Duncan observes, civil society actors in the CFS are sometimes better prepared compared to some member states: 'In the CFS, governments are often represented by UN diplomats skilled in negotiation and politics, however, less knowledgeable on technical issues, and thus often heavily reliant on technical civil servants working back in their capitals' (Duncan, 2014: 159).

Government representatives interviewed for this work repeatedly acknowledged that they find the civil society group to be very well prepared ahead of meetings to ensure consistent, informed and articulate interventions. One government representative from a European country expressed in an interview after a CFS Session how she thought that governments in many ways could learn from how members of the CSM get organised:

> Sometimes it is more effective for me to have meetings with civil society actors than with my own delegation or with the EU delegation. It is my impression that many governments admire how the civil society group works and contributes to the CFS. I must admit that we are not always good at talking to each other. We assume that we know each other's positions. The CSM representatives often remind us about the sensitivity of the issues we are dealing with here.
>
> (Government representative from a European country, CFS Annual Session, Rome, 16.10.2014)

Some government representatives from smaller country delegations to the UN interviewed for this research stated that the CSM group bring issues to the agenda that would perhaps not have been pushed for without active civil society actors in the room. In official discourses and formal interviews, governments often stress how they benefit from the substantive inputs civil society actors bring to the UN. For instance, one European representative explained in an interview that civil society's intellectual inputs often contribute to better policy outcomes. Smaller and less resource-strong country delegations to the CFS do not always have a clear position on issues and interventions. Data from fieldwork conducted in the CFS arena show that positions and 'evidence' presented by the CSM group sometimes can help smaller countries to take a standpoint and intervene in UN sessions:

> Civil society actors do not only help build consensus, they also improve the quality of international decision-making by bringing in views from diverse actors of society. Civil society groups also sometimes encourage smaller country delegations to participate more actively in the debate. This makes a more informed foundation for decision-making.
>
> (UN representative from a Latin American country, CFS Annual Session, Rome, 17.10.2014)

During CFS sessions, country delegates to the UN often address the civil society group and ask them to share their statements and advise governments on different positions:

> Previously, governments did not want to spend time talking to and with us. Now they come to us. They know that we engage in dialogue with segments of society that they do not reach. Peoples' organisations play a crucial part at all levels because they report on the realities in the field that governments do not know about. They need information from the real world.
>
> (South East Asian, LVC member, CFS42 Annual Session Rome, 13.10.2014)

As peasant leaders of grassroots constituencies bring their personal observations 'of the reality on the ground', they contest statements presenting the objective claims of governments: 'With us in the room they cannot just talk wonderfully about what they do at home. Governments are not acting alone anymore. We hold them accountable' (European LVC member, personal conversation, Rome, 20.10.2013). This is an example of how activists in the CSM actively take on a role as 'watchdogs' (Karns and Mingst, 2010: 467).

Interviews conducted in the CFS affirm that government representatives are often well aware that civil society actors engaged in CSM are embedded in a much larger network of civil society actors, ready to go public and mobilise if they find that governments act against the interests of those the CSM actors often defend as the 'the interest of the public' (Observations, CFS-rai negotiations, 2014). As government officials strive to avoid becoming targeted by civil society criticisms and objects of popular protests (Keck and Sikkink, 1998; Smith, 2008) the direct engagement of those most affected by food security and their alliances in the CFS processes clearly brings new dynamics to the policy arena. Yet, whereas government representatives often praise the participation of civil society in the CFS processes, they are not always as open towards antagonistic debates as they claim to be. This aspect of the policy deliberations and negotiation will be further explored in the following section.

b A dissenting voice and disclosure of conflicts

The intensive negotiations over responsible investments in agriculture and food systems were not finalised during the one-week session planned for negotiations of the CFS-rai in May 2014.[24] Bringing together different (and for some states, vested) interests of investments explains why the inclusive negotiations of this political 'hot issue' appeared to be both harsh and conflict-ridden. The negotiations revealed that, while social movements engage in the CFS as 'full participants' to win support for their proposals, they also take on the role of adversaries and challengers. This means that they are not afraid of going into open confrontation if they find that governments' decisions are harmful for those most affected by food insecurity.

Despite the strong pressure exerted on the civil society group to support the final output of the negotiations, the members of the Civil Society Mechanism decided not to applaud the final CFS-rai document. Instead, the CSM group openly expressed the discontent and concerns related to how the CFS-rai will be used in a statement presented by an LVC member during the CFS41, on 15 October 2014.[25] Besides the collective CSM statement, LVC openly criticised the CFS-rai document in their own press release for 'not containing sufficient safeguards to stop land grabbing and other destructive actions by private capital and complicit governments…The majority of governments remain blind to the challenges of global food security' (LVC, 2014c). One LVC member explained how the civil society group, at the end of the CFS-rai process, took on a clear position as 'the dissenting voice' (European, LVC member, Conversation, Rome, 16.10.2013). In this regard, the CFS-rai process demonstrated the uniqueness of the CSM methodology: namely, how civil society actors with no responsibility or obligation to vote, can remain with 'their hands free' to criticise the CFS output that they do not agree with or are concerned about.

For LVC, a radical social movement driven by social conflict and a struggle for food sovereignty, it is important that the CFS arena remains an antagonistic arena with room for disagreement, conflict and confrontational politics. As one LVC member emphasised during the CFS-rai negotiations: 'The richness of global debates is when there is room for conflict. We must stand firm on our right to show that we have different opinions on how to shape this world' (Latin American LVC member, CFS-rai negotiations, Rome, 18.05.2014). However, while the CFS is an inclusive forum of different stakeholders wherein different positions are stated and disputed, the consensus-driven logic of finding a 'balanced compromise' can be highly problematic for radical social movements, as this is seen as a way to avoid addressing existing conflicts and power imbalances (LVC members, CFS Annual Session, Rome, 14.10.2013).

Fieldwork observations in the CFS arena demonstrate that some activists in the CSM in particular find this tendency problematic, as it is seen as diluting the very different mandates held by different 'stakeholders' engaged in the CFS. Consequently, activists within the CSM group sometimes need to remind the CFS chair, and other stakeholders in the CFS arena, that the CSM represents 11 different constituencies of civil society including the 'voices of the honest work of millions of peasants and farmers around the world' and that these mandates should not be given the same weight as those representing private economic interests. One Latin American LVC member explained in an interview that he found that the interpretation of 'fairness' in practice means sustaining the dominant industrial agribusiness paradigm and ignoring the deep conflicts of society (Latin American LVC member, interview, CFS Annual session, Rome, 10.10.2014).

While government representatives to the CFS often invite input from civil society, observations of and interviews with government representatives show that in practice they do not always applaud the way civil society representatives bring their proposals to the CFS table. Fieldwork observations reveal several examples of how some government representatives get provoked by the way civil society groups behave in the policy arena. For instance, during the CFS40 Annual Session in 2013, civil activists from the CSM group fought fiercely to put agroecology on the CFS agenda. As the civil society group members continued to raise their flag and intervene and present their arguments on why the CFS should ask the HLPE to prepare a report on this subject, some governments showed their frustration as the session dragged on. One European UN representative felt so provoked that he left the room in protest. When he passed me on his way out of the negotiations room he expressed that he 'did not have time for the CSM group dragging on the session' (Field notes, Plenary debate during the CFS41 Annual Session, 15.10.2014). Interestingly, these words came from the very same UN diplomat who, in an interview the day before, eagerly praised the value of giving civil society right to participate fully in the CFS.

Another observation was during a CFS Session where a Latin American government representative got so angry that he jumped out of his chair with his country flag waving in the air. The UN diplomat expressed in a rather aggressive tone that the civil society actors in the arena are not 'members' of the CFS but 'participants' and that civil society actors should not have the final word in CFS sessions (Field Observation, CFS Annual Session, Plenary Session, Rome, 16.10.2013). Later that day, I got the chance to talk to this diplomat. He explained his strong reaction to the interventions of the civil society actors in the CFS Plenary session as following: 'You must understand that in our countries we have our own legislations. The problem is that our friends in civil society often tend to confuse the CFS with a human rights binding body. It is not, and it should not be' (Latin American UN diplomat, Rome, 25.10.2013).

The latter example illustrates the at times heated political debate within the CFS and divergent views about the role of different actors as well as the very mandate of CFS itself. This leads us to the next section discussing the role of the CFS within the broader architecture of global food governance.

d A struggle for democratic control

The introduction of this book presented how CFS was born in the midst of a struggle over its political significance. It is worth recalling that the reform of the CFS in 2009 was conducted at the same time as another proposal for a global level coordination mechanism – the Global Partnership – was being promoted. Whereas some governments strive to downplay the political weight of the reformed committee (Duncan, 2015; Brem-Wilson, 2011) civil society actors engaged in CFS processes struggled fiercely to reform the CFS from a 'talk shop' into an action-oriented Committee (McKeon, 2015: 105). A LVC member expressed during the annual civil society forum of the CFS 41, in an evaluating of the CFS: 'For us, the CFS should never be reduced to a debate club. The reformed CFS is a political arena. It is the major international policy forum for decision-making on agricultural and food issues' (LVC member, interview, Rome, 21.10.2012).

While LVC and other of its allies within the food sovereignty movement would like to see the CFS as the single authoritative body in the global governance of food and agriculture, some governments continue to resist the incursion of CFS into specific policy areas, and especially trade (Observation, CFS-rai negotiations, Rome, 2015). In this regard, the LVC's choice of engaging with and within selected UN institutions, in particular those addressing issues of food and agriculture, is a historical struggle to take the control of food and agriculture out of instances that clearly impose a neoliberal trade agenda, and especially the WTO (Rossett and Martínez, 2010; Desmarais, 2007), into a more transparent and democratic forum for global policy-making.

The right of peasants and small farmers to have direct impact on agricultural policies at all levels of policy-making has been a central component of food sovereignty from its origin with 'democratic control' as one of the key pillars of food sovereignty: 'The United Nations and related organisations will have to undergo a process of democratisation to enable this to become a reality. Everyone has the right to honest, accurate information and open and democratic decision-making' (LVC, 1996; see also Brem-Wilson, 2015).

The participation of LVC in the CFS is part of the larger struggle for 'a voice, legitimacy and recognition at the global scale' (Desmarais, 2007: 197). A LVC member emphasized, during a world social forum event, how the CFS is an important achievement for citizens seeking to democratise global governance of food.

> Before, we were not allowed to take part in these kinds of meetings. Governments have sought to limit people's access by attempting to shift major policy decisions to take place outside the UN and into more exclusive global financial arenas where small-food providers and other civil society organisations have no control at all. We may not win all struggles in the CFS, but we keep on saying that all decision-making processes must be inclusive. Otherwise, there will not be counter-responses to those supporting the corporate interests.
>
> (Latin-American LVC members, workshop organised by the Civil Society Mechanism, World Social Forum, Tunis, 25.03.2013)

LVC's engagement in the CFS can be seen as a struggle, not only to influence the output of the policy process but a fundamental struggle over the right to decide the meaning of what key concepts such as democracy mean. As one social movement activist from the Global South expressed in an interview after a CFS Annual Session:

> I sometimes wish we could sit down and have an honest talk about democracy, representativity, legitimacy and citizenship. What does it mean to elect governments? We vote, but do they respond to our claims? Do

> they protect the interest of citizens? Which interests do they serve? For us being here, it is not only about added value of civil society participation. It is about changing the whole way of thinking, about who has the right to contribute to the production of society. It is about going beyond dialogue and consultation and address those responsible for affecting the lives of people whether these decisions are taken at a global, national or community level.
>
> (Social movement representative form the Global South, personal conversation, Rome, 16.10.2013)

Returning to the concept of 'cultural politics' presented by Alvarez et al. (1998) social movements are not working for 'inclusion' in existing political structures and the dominant culture. Instead, social movements' activism is often directed towards structural transformation of the very political order in which they operate:

> they [social movements] seek to democratise[26] sites and structures of power, limit the power of those sites and structures, and their vision for social change often encompasses developing a political culture that is based on the principles of transforming society as a whole.
>
> (Alvarez et al., 1998: 8)

From this perspective LVC's engagement in the CFS can be seen as part of a larger struggle over the right to contribute, to produce and shape society with different forms of knowledge. An LVC member explained during a CSM session that peoples' participation in global policy processes is important for what politics means: 'Politics is too important to be left to politics. Politics is more than a compromise: it is a struggle over values and worldviews' (Latin American LVC member, personal conversation, Rome, 16.10.2014). These antagonistic elements are important for understanding the rationale for social movements' engagement in the UN Committee. For peasants' representatives, this is a simultaneous act of contesting and seeking to engage strategically to negotiate within the institutions with the aim of gaining support for their struggles.

These observations of social movements within a UN arena confirm that the CFS is a political battlefield where social movement actors are acting simultaneously as 'participants' and 'challengers': they are both shaping and contesting the very political order in which they operate. The attempt to win support for alternative worldviews and proposals to shape the future of global food system, remains, however, an ongoing and intensified struggle.

> We must always remember that we are not the only players here. Others seek to take this space and other spaces. After being here, I really understand why we cannot let others decide for us here anymore. It is important

> to be here, because as we say in La Vía Campesina: No Decisions About Us Without Us. We must keep filling this space with the voices and demands of peasants.
>
> (European LVC member, CSM evaluation meeting, CFS41 Annual Session, Rome, 24.10.2014)

5 Conclusion

This chapter has explored some of the dynamics in the reformed CFS as a political battlefield, where deliberation, negotiations and confrontation take place. It has shown how peasants in the CFS fiercely seek to change the terms of the debate by politicising the arena in the ongoing struggle over meaning-making (Kurzman, 2011) and political manoeuvring.

Food sovereignty seems increasingly to be the political frame that brings a diversity of civil society actors together to shape a coherent vision for the future of food security built on the rights of peoples, communities, and countries to define their own food systems and agricultural policies. Even though food sovereignty is not integrated in official UN text, civil society actors continue to find ways to reframe the terms of debate to support the pillars of food sovereignty. Interviews conducted during international meetings reveal that social movement activists are often aware that the premise for their engagement in the UN arena dominated by governmental actors is a constant struggle to act strategically and do political maneuvering.

LVC is an example of a movement actively seeking to negotiate its engagement and shape the arena developing its tactics and strategies in the attempt to do political manoeuvre, shape the agenda, and assert the group identity as peasants. The ongoing struggle in this policy arena compel us to be careful not to 'measure' the success of social movements from too narrow an understanding of effectiveness. As both 'participants' and 'challengers' social movement activists give strong impetus to the struggle over the global food system, simultaneously shaping the ongoing policy processes and challenging the very foundation for how global policy-making takes place. Whichever direction the struggle will take, with the entry of new actors in transnational policy-making, the struggle of the global food system has been expanded, deepened and intensified.

Notes

1 These negotiations culminated with two rounds in May and in August 2014.
2 For instance, the two-year processes of inclusive consultation and negotiations of the Voluntary Guidelines on the Responsible Governance of Tenure of Land, Fisheries and Forests in the Context of National Food Security (VGGTs) where responsible investments were identified as a particularly contentious issue (Duncan, 2015). The CFS negotiated guidelines are available here: http://www.fao.org/fsnforum/news/new-voluntary-guidelines-responsible-governance-tenure-land-fisheries-and-forests-endorsed.

3 I here refer to the civil society alliances built around mobilisation against global land grabbing as well as the network of the International Planning Committee for Food Sovereignty, supported by NGOs like FIAN (Food First Information and Action Network) and GRAIN (see https://www.grain.org/). The network argued that the PRAI principles were a 'Trojan horse' to promote an agribusiness model that threatens food security of the local population – who are dependent on their land and their crops to achieve food security – and ultimately undermines local food sovereignty (Gaarde and Hoegh-Jeppesen, 2011: 50).
4 Field work research conducted within the Network of Farmers' and Agricultural Producers' Organisations of West Africa (ROPPA) in 2010 witnessed the dense networking against land grabbing both 'horizontally' among social movements around the world as well as 'vertically' with some governments and representatives from some international institutions and agencies (FAO, OECD).
5 At this time, the formulation of guidelines on land tenure governance had already been mobilised to the negotiation table at the CFS. In this regard, Duncan (2014) has presented how Voluntary Guidelines on the Responsible Governance of Tenure of Land and Other Natural Resources – VGGT (VGGT) shifted from a technical FAO process to a political nature of the CFS leading to increased influence and interest (2014: 173).
6 Expressions used by European LVC member in a preparation meeting for the CFS-rai negotiations (observation, Rome, 24.04.2014).
7 For instance, McKeon has argued that building food sovereignty has been a counter-frame to the term food security

> framed in neo-liberal terms as increasing productivity per plant/animal, making food available through formal markets and imports, counting on economic growth to improve incomes and employment, and – along the way – reducing pressure for agrarian reform. In this logic, peasant-based production was backward and inefficient.
>
> (McKeon, 2015: 77; see also Wittman, 2009 and Desmarais, 2007)

8 While the CFS has recognised the overwhelming role of small-scale producers in feeding the majority of the world, and the HLPE, also stated in their report on smallholder investments (HLPE, 2013: 34) that 'smallholders are the main investors in agriculture', the CFS-rai principles, undermined that statement by affirming that they are the main investors only in their own agriculture (McMichael and Müller, 2014: 5).
9 For instance, on several occasions during the CFS-rai negotiations, the CSM group emphasised the extraterritorial obligations of states to secure the Right to Food and to regulate and make private companies accountable for their operations abroad, as a way to uphold international human rights obligations over trade and investment regimes (Participant observations, CFS-rai negotiations, Rome, 20.05.2014).
10 I borrowed this title from Birgit Müller and Catherine Neveu (2002).
11 Publicly stating that someone has acted in a bad or illegal way, for instance violating rights (Keck and Sikkink, 1998: 6).
12 A number of scholars have observed how governments often seek to move the debate on issues of technicalities as an attempt to render the debate 'anti-political' (Müller, 2013; Ferguson, 1994).
13 Expression used by an NGO member to the CSM (Observation, CFS-rai negotiations, Rome, 04.08.2014).
14 Group efforts to reassert the claims when someone else challenges or violates them (Tilly, 1978: 144)
15 FPIC stands for Free, Prior and Informed Consent, and is strongly advanced by LVC and other social movements. FPIC is today considered as an international

human rights standard that derives from the collective rights of indigenous peoples to self-determination and to their lands, territories and other properties. See FAO, 2012a.

16 Mainly allies such as researchers, but also occasionally influential actors such as the UN Rapporteur to the Right the Food (Observation, CFS40 Annual Session, 2014).

17 This was also what Duncan found in her study (2015) of the civil society group in an earlier CFS process (the Voluntary Guidelines on the Responsible Governance of Tenure of Land, Fisheries and Forests in the Context of National Food Security, negotiated in 2011–2011).

18 Throughout this Symposium, 300 participants from academia, farmers' organisations, government representatives and FAO experts exchanged views and experiences with agroecology. An LVC member expressed how LVC strongly strove to emphasise the political dimensions of agroecology (Latin American Member, LVC, interview 10.10.2014).

19 As Rosset and Martínez state: 'The right to continue to exist as such has been the point of unity of peasants in the world' (2010: 149).

20 Other observers have stated how high input agriculture and 'technological scientific knowledge' historically have dominated the 'metis' of local farmers (Desmarais, 2007: 28; Scott, 1985; McMichael, 2006; McKeon, 2015). In other words, it is a struggle over the right to shape society from different sources of knowledge.

21 The international Nyéléni Newsletters are written by citizens engaged in the international movement for food sovereignty. See: http://www.nyeleni.org/spip.php?page=NWarticle.en&id_article=372

22 http://www.fao.org/cfs/cfs-hlpe/en/

23 Data from field research conducted at the CFS, but also at the UN headquarters, reveal how a multitude of actors – including government and UN staff from other UN agencies – recognise the CFS for its inclusive and consensus-seeking approach. Olivier de Schutter, at that time the United Nations Special Rapporteur on the Right to Food, also expressed this point, (email correspondence, 07.05.2013).

24 Despite the strong pressure from the CFS managers to finish within one week including night sessions, the contentious character of the issue led to a second round of negotiations in August 2014, where the main part of the negotiation text was finished. The final CFS-rai was endorsed during the CFS41 Annual Session without the applause of civil society.

25 http://www.csm4cfs.org/news/cfs41_principles_for_responsible_agricultural_investments.13/ Accessed 1 February 2016.

26 Social movement activists often conceptualise democracy as both a process and an end (see e.g. Doucet, 2008: 20–30).

5 From the plough to the UN negotiating table

The previous chapter showed how LVC has played a key role in carving out an autonomous civil society space within the UN Committee of World Food Security that was reformed in 2009. We saw how one of the central components of the CFS reform was to grant self-organised non-governmental actors 'full participation'[1] in the processes leading to policy-making in an UN intergovernmental committee, previously restricted to governments. With an interest to explore *agency* (the capacity to act),[2] this chapter explores how social movement activists – mainly those representing food producer constituencies – experience their engagement in the CFS arena. The context of this chapter is the debate around the aspirations for inclusiveness of the CFS, and the challenge related to translating 'formal' into 'substantive participation' (Brem-Wilson, 2015). The chapter argues that social movements – despite the immense challenges faced by peasants engaging in a world dominated by state bureaucracy and expressed in the languages, codes and norms of the UN arena – do not automatically follow what I, in the introduction of this book, called the 'classical path' of institutionalisation, leaving little room for innovation and agency.[3]

The chapter argues that the autonomous Civil Society Mechanism (CSM), as a collective space, functions to a high degree as a laboratory for developing innovative ways to support the agency of 'newcomers' to the CFS. In particular, the chapter shows how committed and skilful interpreters play a fundamental role in overcoming some of the structural barriers related to language and translations. The chapter ends with a call for more research to further explore the unfolding structure-agency dynamics arising from social movements' participation in this transnational policy space.

1 Finding the way to the negotiation table

First of all, entering a formal UN intergovernmental arena like the CFS may be a rather bewildering experience for peasants, as well as other 'newcomers', who are not experienced in engaging in intergovernmental meetings. Participating in a formal UN arena means gaining knowledge of, and familiarising yourself with, the culture of UN diplomacy, with distinctive codes, modes of communication and undecipherable language inherited in the meaning-structures (Holzscheiter, 2010: 734).

Participation in the UN means that social movements have to engage in an arena dominated by UN diplomats and UN policy professionals, where technical information, acronyms, and formality dominate the culture (McKeon, 2009a: 63). The sterile bureaucratic and formal atmosphere the social actors face when they enter UN intergovernmental negotiations contrasts in many ways with the very dynamic energised, symbolic and grounded atmosphere of the peasant movements' own grassroots spaces.

For instance, during the Jakarta meeting, every morning session started with a 'mistica' where people stand up, move their bodies and gather energy for a day of meetings. 'Misticas' are spiritual and cultural acts where members can express and connect viscerally with their struggle, using songs, poetry, dance, or theatre (Desmarais, 2007: 184). People entered into the hall equipped with flags, music instruments, traditional drinks, food and seeds and other symbols of the peasant struggle and strong symbols of their bond to each other and to earth, reminding the peasants about the struggles they are facing everyday all over the world. As other scholars have stated (McKeon, 2009; Brem-Wilson, 2015) these differences between the communicative opportunities in the UN and the movement language means that some officials become conscious of not quite understanding the movement until they have participated in the movement's native arena.

During my time conducting research in UN arenas, I often remembered my own first experiences of participating in UN meetings. Based on my memories of the effort I made to decrypt the diplomatic codes of participation in cumbersome UN meetings, I continued to ask myself whether we could expect peasants to feel comfortable in a UN setting. Engaging 'fully' in an intergovernmental arena means that member participants must follow the working logic, the formality, speed and rhythm of a UN arena (Duncan, 2014; Brem-Wilson, 2015). Following how the UN negotiations play out can be rather confusing. For instance, when member states raise their country signs during CFS negotiations to suggest a change to the text, the changes are put in different colours in the text. As experienced by an indigenous peoples' spokesperson to the CFS: 'The text on the screen is changing all the time, what does this mean when the colours change, does it mean that our text proposals are lost?' (Field observations, Indigenous Peoples' representative, CFS session, 20.05.2013).

A LVC representative described that engaging in the CFS can feel like an overload of complex information. The UN has internal technical working modes, logics, and protocols that all newcomers must break through: 'You need to find your way and break down the information you receive. I used the first year to understand the complex structures and the dynamics of 'how this big machine works'' (English speaking LVC member, personal conversation, Rome, 16.10.2013). Besides the range of personal challenges for social movements to get closer to the idea of 'full' participation of civil society actors, the issue of language remains a main structural barrier for social movements.

a UN language is power

A complete understanding of the predominant language in global decision-making processes is a key prerequisite to full and equal participation in the UN arena (Brem-Wilson, 2014: 13). A native English-speaking CSM member, participating in a global UN meeting for the first time, explained how technical UN language is a barrier to full participation for many newcomers in the CFS:

> You easily get lost in the language here. We are drowning in acronyms: what is the difference between PRAI/RAI?[4] Even for me as a native English speaker I'm struggling to follow the language. We need a UN dictionary for social movements here...There are so many paragraphs in the text and sometimes they say nothing to us. It is a very abstract and very fuzzy language they use here. We keep going back and forth in the text, the sentences are so long and detailed and it is difficult to follow...What does 'the spirit of consent' mean? We are told that consensus means that we should agree if we could live with the decision. I can live without one leg, but I would certainly prefer not to.
>
> (CSM member, newcomer to the CFS, conversation during the CFS49, Annual, 18.10.2013)

This newcomer to the CFS revealed the issue of language and the sense of confusion food producers or food activists may feel whilst seeking to decode the UN language when they enter the CFS room for the first time. In order to fully engage in the CFS arena, participants must speak one of FAO's six official languages,[5] for which interpretation in official meetings is provided. Data from observations in the CFS arena show that social movement representatives often remind the other stakeholders about the power disparities inherent in language. As Brem-Wilson points out, the FAO itself has indeed previously noted an informal tendency, within the various different fora of its work, to default to English, with the arising issues of inclusion and exclusion (Brem-Wilson, 2015: 13).

In an interview a Spanish-speaking LVC leader explained that he felt that native speakers more easily gain the attention of the audience when speaking in their respective mother tongue, as many substantive as well as rhetorical aspects may be lost in translation. He went on to present the example that UN translators often translate 'peasant' into 'farmers' which is problematic for LVC delegates who often hold a strong value to their politicised peasant identity (Borras, 2004; Desmarais, 2007). The problem here is partly that international institutions like FAO have not developed glossaries that perfectly correspond to the political terms of a social movement.[6]

For instance, a LVC interpreter explained to me in an interview that the problem is that 'peasant' is generally seen as backwards or directly negative. One of the interpreters to the CSM explained that outside the circles of the food sovereignty movement, the word 'peasant' often has a negative connotation.

Citing the *Oxford Dictionary* (2014), a peasant is 'A poor farmer of low social status who owns or rents a small piece of land for cultivation; informal, an ignorant, rude, or unsophisticated person'. The official UN interpreters are trained to follow the dominant discourses and codes within a UN setting (CSM Interpreter, Conversation, 30.10.2014). This may be one of the reasons why it remains of utmost importance for LVC to work with its 'own interpreters', who are aware of the political significance of the terms often used by social movements. As the next section explains, committed interpreters play an important role in overcoming language conflicts and 'intercultural misunderstandings'[7] (Doerr, 2012). As scholars of linguistic anthropology (e.g. Fairclough, 1992) remind us, language is embedded with a subtle sense of the complex interworking of meaning, style and social positioning. One native English speaker reflected on the barriers of 'full' social movement participation in the CFS when I sat next to him at a CFS annual session.

> I am a native English speaker and for me there is still a language of exclusion here. It is not about being an English speaker or not. The problem is much bigger and cannot be solved by translations. The UN language is one of politics and power. When you go to the CFS for first time you easily get lost in the language spoken. We all talk in acronyms – but how can that be translated in the field? I am member of the CSM, IPC, FAO, FN…it's like a jungle. We need an acronym handbook for social movements to be able to engage here.
>
> (Personal conversation, farmer and CSM participant, Rome, 15.10.2013)

> It is so technical. Sometimes I read a sentence over and over again and never get to understand it. The UN should simplify the language spoken here, so it is comprehensible for citizens all over the world.
>
> (Asian LVC member, CFS Annual, 12.10.2013)

As we saw in Chapter 3, the reform of the CFS arena in 2009 led to an opening up to the inclusion of actors from historically marginalised sectors of society, such as peasants, indigenous peoples, fisher folks, pastoralists etc. Today they engage in CFS deliberations and negotiations; however, some of these groups are not always used to – or comfortable with – speaking about their struggles in the colonial languages or UN languages. Post-development scholars remind us that groups that are under 'global colonial rule' (Escobar, 2004) often engage in an epistemological struggle against Eurocentric values inherited from colonialism (Escobar, 1995; Santos, 2012). According to Santos these groups often hold strong holistic and spiritual values and seek to re-interpret common terms like human rights, democracy and development to give way to dignity, respect, territory, self-government, good life, and mother earth (2004: 48). This meeting between different world visions, Santos argues, calls for 'intercultural translation before they can be understood and appreciated' (ibid.).

At the end of this chapter, I return to how skilful interpreters are important for the group to work efficiently across languages and cultures despite language diversity (Doerr, 2012). The following section presents some of the main tensions faced by social movements seeking to gain their autonomy in the CFS arena, where the efficiency logic, working culture and communication modes of the UN intergovernmental world prevail.

b Different meaning of 'efficiency'

Chapter 2 showed how LVC organises itself as a people's organisation, relies on a decentralised structure, and holds strong values of inclusion. It means that LVC delegates to international meetings (ideally) must take the time to build positions of broad consultation among its members. However, inclusion is often presented in the social movement literature to be in conflict with the need for efficiency in meetings: 'Efforts to improve internal democracy and openness of decision-making processes slow down and make the answers to problems in a changing context more complex' (Pleyers, 2012: 178, citing Gaventa, 2004: 312). Observing the dynamics in the CFS arena shows that the very meaning of 'efficient' is at stake. It sometimes means different things to different actors engaged in the CFS. The international governmental logic is often strongly outcome-focused and driven towards reaching a consensus between all stakeholders.

The CFS-rai negotiations (2014) revealed these different logics of efficiency from the outset. In the opening session of the first round of negotiations of investments in agriculture, the CFS Chair initiated the negotiations by presenting a picture on the monitor screen in the negotiation room showing a group of hikers climbing towards the top of a mountain: 'This is where we need to go. We need to support each other by rope to climb to the top. We need to give in and we need to keep focus' (CFS Chair, Opening of the CFS-rai negotiations, Rome, 17.05 2014).

Reaching a final document 'that we all can live with' was from the beginning a strong imperative from the UN/CFS logic and showed how 'UN efficiency' often means leading stakeholders efficiently towards decision making and sticking to a tight time schedule.[8] Social movement activists find it difficult to acclimatise to the CFS efficiency logic and the institutional semantics that revolve around the discourse of 'the world is waiting for you'.[9] Moreover, they are reminded of the CFS message 'we are walking together', which puts them under pressure to find common ground. While UN diplomats may see it as a failure of their professional diplomatic skills if there is no consensus reached at the end of the day,[10] social movement participants to the CFS often remind other stakeholders in the CFS that they need adequate time to consult with their constituencies: 'We cannot clear a position by making a short phone call back to capital. We have 172 capitals to refer back to' (Representative of LVC in the CFS plenary session, Rome, April 2013).

The problem with UN efficiency was strongly articulated by Paul Nicholson, a Basque farmer and one of the founding members of LVC, when I interviewed him about the challenges of the movements to engage with the UN: 'If efficiency means to go fast, forcing consensus and voting processes, it means institutionalised pre-determinacy. It fails and this is not democratic. If this is the logic, I do defend inefficiency much more than I defend efficiency' (Paul Nicholson, Skype interview, 06.06.2013).

The difference between the 'social movement effectiveness' versus 'UN efficiency' was described to me by a food sovereignty activist engaged in the CSM from an organisation applying for LVC membership, in the following way:

> Bringing the people with you in global meetings is very important. In fact, I would say that it is impossible to achieve food sovereignty from the top down. This means, being organisationally efficient rather than effective. That is often counter-productive.
>
> (CSM participant, CFS session, Rome, 14.10 2014)

These examples demonstrate how mobilised peasants, who are delegated to international negotiations, seek to found their engagement on strongly held principles of inclusion while taking part in international negotiations. It illustrates how 'working slowly towards a UN outcome' may the most efficient way to move ahead for the internal functioning of social movements. As one LVC member expressed in a reflection about UN engagement, he did not feel that such questions were sufficiently addressed within the UN arena.

> I sometimes feel that we use too much energy discussing internally how we can work best and efficiently vis-à-vis our constituencies and the responsibilities we have [at the CFS]. But what about asking how to make the UN system and policy processes more efficient for social movements?
>
> (European LVC member, interview, CFS session, Rome, 13.10.2014)

The following section discusses the dominant communication mode and how it has an impact on the way agency plays out.

c The prevailing mode of communication in a UN arena

For rural constituencies, presenting a short and concise message in the UN arena populated with hundreds of high-level diplomats, international 'experts' and ambassadors from around the world can be intimidating for those who are not used to speaking publicly or in front of a larger audience. This may be particularly the case for those that do not have translations provided to them to speak in their mother tongue. Brem-Wilson theorises how, in order to participate effectively in the CFS as a 'discursive arena', the interlocutor must possess a degree of psychological comfort with the dynamics of participation in the arena. This includes having confidence in their own right to speak, and to not be intimated by the status of the others (Brem-Wilson, 2015: 14, citing Gaventa, 2004).

Data from fieldwork conducted in the CSM show that social movements actors seeking to influence intergovernmental UN negotiations are often aware that it can be advantageous if social movements speak in accordance with what Brem-Wilson calls 'the prevailing mode of communication' in the arena (2015: 13). For instance, as one indigenous youth leader explained to me during the CFS-rai negotiations, it would be more efficient if she were to convey the message in 'UN language': 'Of course I am talking here like I talk at home. I seek to formulate my arguments so that governments in this room actually listen to me' (Indigenous youth leader, Rome, CFS-rai session, 20.05.2014).

In his work exploring rural movements' engagement in CFS during the first years of the reform process, Brem-Wilson suggests how one could imagine civil society actors being allowed to introduce new rules or norms in the CFS arena. For instance, a norm could be one that fully recognises each member to speak in his or her 'natural' mode of communication – despite certain concerns of 'cost' related to loss of (UN) effectiveness (2010: 200). Other scholars also critically address the issues of communication in the CFS arena. For instance, Duncan and Barling affirm:

> There is a risk that the participatory nature can become 'overly cognitivist or rationalistic and thus insufficiently egalitarian' by favouring the 'educated and the dispassionate' and excluding 'the many ways that many people communicate reasons outside of argumentation and formal debate, such as testimony, rhetoric, symbolic disruption, storytelling and cultural- and gender-specific styles of communication'.
>
> (Duncan and Barling, 2012: 157 citing Bohman, 1999: 410)

While social movement activists of the CSM group seek to address the power gap inherent in the CFS arena due to the English-dominated and technical UN language, trained government representatives and resourceful NGO staff (often educated at elite universities and trained in values of Western world-views) have an intrinsic advantage and can better adjust to the working culture of the UN.

2 The CSM group: strategic team negotiations

a A laboratory to bring social movements into political leadership

When civil social actors arrive in Rome to participate in the CFS they enter the UN arena with different degrees of personal experiences. Some are more comfortable speaking in a language other than their mother tongue and talking into a microphone, while others find it more difficult to adapt to the working mode and codes of a formal UN setting. The building of the CSM as a collective civil society space means that CSM members do not act in the UN arena alone. As presented in Chapter 3, the core vision of the CSM is that those most affected by food insecurity and malnutrition must be agents of their own development and change (CSM, 2010a: paragraph 10).

Members of the CSM group are aware that agency is linked to a human capacity to take on political leadership. This means that members of the CSM group seek to overcome a number of barriers for social movements' 'full' participation.

In the case of LVC, the peasant delegation to the CFS normally arrives in Rome some days ahead of the international meeting to prepare for the specific process. These preparatory meetings are normally followed by one or two days of intense preparation meetings within the broader civil society group. During such internal meetings, technical understanding is improved, and actors have the possibility to learn from others in the meetings, prepare their interventions, and draft the text to be deliberated/negotiated in the CFS.

Fieldwork observations show that the CSM is in many ways a collective project where more experienced members of the group help to support newcomers when they arrive in the UN for the first time. This includes supporting newcomers to the CFS by 'breaking down' concrete UN texts and sharing experiences. This both helps to demystify how the UN arena works and to support the agency of those recently arrived at the CSM or those with fewer prerequisites to participate in the CFS processes. When negotiations in the CFS plenary gain speed, members of the CSM support each other with comments on notes on papers circulating back and forth between members in the CSM group.

b Focus process

Besides the effort to influence the concrete outcome, the civil society actors of the CSM unite around a strong focus on *process*: exchange of experiences, knowledge and different resources. This was, for instance, observed in a CSM meeting when an LVC 'newcomer' to the CFS processes during a CSM preparatory meeting was asked by one of the CSM secretariat members if he would like to try to chair the meeting. The LVC delegate, looking rather surprised, responded that 'he was not sure if he could do that' and that he was basically 'just trying to find his way from one meeting room to another in the bewildering FAO building' (Field observation, CSM meeting, Rome, 10.10.2014). A few days later the member accepted to chair the meeting. In a personal conversation, he narrated how he considered it to be a 'personal triumph': 'I was very surprised and very nervous, but after this experience I believe more in myself. I feel stronger, and I don't feel that I am acting alone' (European LVC member, newcomer to the CFS, Rome, 13.10.2014).

Members of the CSM group often refer to this as a personally challenging task; they also refer to the psychological support they receive from others in the CSM group.

> I am very nervous when I intervene in the CFS plenary but I am also confident because I know that I am not alone, I feel backed up by a team.

> I feel that we are learning how to act here every day. In the beginning it feels like we use a lot of energy, but in the long run we will put pressure on governments and we must continue to do this – not only here, but even more importantly, to bring the pressure home.
>
> (European LVC member, newcomer to the CFS, interview, Rome, 09.10.2013)

The collective engagement to the CFS can give an increased sense of agency. This was also registered during the last round of CFS-rai negotiations. After two years of deliberations and negotiations, governments succeeded in getting a WTO reference into the final document. This was a great dissatisfaction for members of the CSM group who fiercely stated that the CFS should adhere to a human rights-based framework and should not be guided by trade interests (Observation, CFS-rai negotiations, Rome, 08.08.2014).

At the end of the session, an LVC member delivered the strong political statement on behalf of the CSM group against the WTO. When the last session of negotiations closed around midnight, I asked the LVC member how he had experienced this whole scenario. Following the civil society group during CFS preparations, negotiations and evolutions from early morning to late night many days in a row, extremely tired myself and expecting the LVC member to be worn out, he replied:

> Without me, it would have been much worse. Governments and their friends from the private sector allies wanted the WTO into the document many times. I told them that the WTO does not have a place in food and agriculture. Corporations and interests that go against small producers and peoples' projects all over the world dominate the WTO. I was a shield against more influence of the WTO.
>
> (Asian LVC member, Rome, 18.08.2014)

The peasant delegate expressing his own empowerment here is in line with Amartya Sen's approach to *agency*, namely as 'an agent or someone who acts and brings about change, whose achievement can be evaluated in terms of his or her own values and objectives' (Sen, 1999). The focus on individuals and collective experiences of social actors is also a key component of a number of contemporary social movement studies (Wieviorka, 2005: 283–310; Pleyers, 2010: 35–57; McDonald, 2006). This aspect of agency is particularly eminent in Touraine's approach to social movements, presenting personal experience as a key component to bringing about societal change: 'it is not universal values, but personal experience of individuals that will bring about mutations in society' (Touraine, 2002: 391).

In the context of social movements' engagement in the UN arena, the agency emerges from strong collective learning experiences that members go through in such processes as well from a strong emphasis on the power behind uniting different types of experiences. A young lady representing the workers in the CSM expressed this in the preparation meeting to the first round of the CFS-rai negotiations:

> We must find other ways to be unbreakable when we move into negotiations with governments. We do not have a rich lobby industry to work for us here. Therefore, we must do our best as a group to gain from the strengths of our diversity. There is more knowledge and wisdom dispersed in a team than in a single mind.
>
> (Field note, CSM preparation meeting for CFS-rai negotiations, Rome, 10.05.2014)

The experiences that members gain from organizing within the CSM group, both during self-organized face-to-face meetings as well as the collective negotiations, boost the self-esteem of individual members and the empowerment of the group. This strong agency (will and capacity to act) is further enhanced by allies and loyal supporters of the CSM, in particular, committed interpreters and academics.

c The role of committed interpreters and academics

One way that the CSM group seeks to support the diversity of 'full participation' of those actors who do not speak the dominant languages, is that the CSM group, when possible, provides its own interpreters to the meetings. These skilled interpreters are playing a vital role to help overcome communications and cultural conflicts within the CSM group.

Ahead of CSM/CFS meetings, the CSM interpreters acquaint themselves carefully with the specific international conventions and texts. Several of these interpreters have been involved in voluntary work for LVC/IPC years before the CFS reform and before the CSM started paying interpreters.[11] When the CSM was designed, it was a demand from LVC to work with interpreters that have volunteered for the movement in previous meetings.[12] The group of dedicated interpreters assists the CSM group in internal meetings and often accompanies the group when they enter CFS meetings. Although there is interpreting provided during the CFS plenary sessions, during negotiations CSM interpreters often sit next to non-English speaking negotiators in the frontline of the negotiations, providing language support to follow the negotiation text on the screen, which is only in English.

One CSM interpreter explained how as part of the interpreters' (and translators') engagement with the CSM they agree to be 'on call'.[13] Data from this research also reveal that interpreters at times work under very difficult sound conditions. For instance, when the CSM group is moving around the FAO building to find a place in the corridors to hold strategic meetings in privacy (Field observations, CFS-rai negotiations, 20.05.2014). Additionally, there are not always 'booths' available for interpreters. Observations reveal various examples of how CSM interpreters have found ways to adjust to using alternative equipment, such as a 'spiders', providing a microphone to the interpreter and several headphones for a small group.[14]

Data from fieldwork conducted in the CSM show that these interpreters are highly committed to supporting the improvement of social movements' engagement in UN meetings. For instance, in an interview a CSM interpreter shared with me a list of reflections identifying possible ways to strengthen social movements' participation in the CSM group, e.g. by using more innovative multi-lingual Skype calls (CSM/LVC interpreter, Skype interview, 30.05.2014).

Interpreters also contribute in supporting the agency of the social movements in the CFS by being available for supplementary language assistance to 'split up'-meetings. When in line with other intergovernmental negotiations, the so-called 'Friends of the Chair' or 'Friends of the Rapporteur meetings' may be established during negotiations in attempts to advance consensus in smaller separate groups. These meetings are mainly in English.

One way the CSM group seeks to make up for the language (and power) gap is to provide their own translators and interpreters to these CFS meetings, where resources are available. In these meetings the CSM group requests CSM interpreters to translate for CSM participants only. This means that non-English speaking UN diplomats must 'live with' some CFS related meetings only taking place in English. As expressed by a CSM interpreter: 'If member states want to have interpretation in all meetings, they must also pay for this' (CSM interpreter, Skype interview, 15.05.2013). Members of the CSM express their gratitude to translators and interpreters who are supporting the functioning and the vision of bringing social movements into the political leadership in the CSM group. As expressed by an LVC member in a CSM evaluation meeting: 'They are much more than interpreters, translators, and technical support, they are our friends and closest allies. Without them the CSM could not happen' (Spanish speaking LVC member, CSM evaluation meeting, 13.10. 2014).

Finally, besides, the contribution from a handful of dedicated interpreters helping both to facilitate and support the agency of particular social movement delegates in the CSM group, the mechanism is supported by a much larger network of people, notably civil society experts and academics educated in human rights. This was, for instance, registered, during the first round of the CFS-rai negotiations, where members of the CSM group connected with allies on a joint Skype chat. This helped CSM working group members to quickly reply to each other's questions and to connect with those civil society members not physically in Rome. It was also a way for experts in the larger network to provide technical expertise to the group from abroad.

These observations reveal that both dedicated interpreters and academics contribute to the collective task force in the CSM.

3 Between unease and human agency

The third section of this chapter reveals that despite a number of personal challenges to engage in the CFS, members of the rural constituencies that are engaged in the CFS often express a strong will and responsibility to 'fill' this global political space with the voices and demands of peasants.

a The compromise

Before and during CFS sessions, LVC delegates, along with other members and participants of the Committee from around the world, spend many hours negotiating, debating and (re-) writing texts. Paragraph by paragraph, words are carefully chosen. Sometimes, even moving a comma leads to heated arguments. However, in the CFS, the resolutions that are discussed for the adoption of text are reduced to small boxes with a synthesis by the time they arrive to the CFS plenary session. The nature of UN processes is strongly directed towards compromise and consensus, which naturally moves all actors away from their original objectives (Barling and Duncan, 2012: 155, citing Mouffe, 2000: 17).

The compromise – the essence of intergovernmental negotiations – can, however, be very difficult for social movement activists to accommodate, in particular as activists are often politically, intellectually and emotionally tied to their standpoints (Duncan, 2014: 149). An LVC member expressed this with regard to the whole 'mindset' of the UN negotiations during his first encounter with the UN world during a CFS session: 'It is difficult for me to see how all these small details in a text are so important. I am used to working on my farm before the sun gets up and then seeing the result of my work when the sun goes down' (European LVC member, 'newcomer', interview, CFS, Rome, 10.10.2013).

The nature and incremental change of deliberations can be difficult to assimilate for peasants as they are used to seeing immediate results of their work 'here and now'. CFS outcome documents can often be far removed from the complex struggles faced by peasants in their fields around the world. While the constituency members use their personal time, energy and effort to reach common proposals and compromises within the civil society group, these may disappear in the following draft version. For instance, fieldwork observations in the CFS reveal that it can be frustrating for social actors, who seek to urge governments to endorse more ambitious goals, to see their proposals being weakened by notions like 'could' or 'being encouraged'. 'It is not only that we discuss all these words again and again and you feel that you are getting nowhere. It is also this feeling that you give in again and again. How much can we give in?' (Latin American, LVC member, Field notes from CSM wrap-up meeting CFS-rai negotiations, 30.05.2014).

Social movement delegates often reiterate the importance of building a radically different food and agricultural model, rather than creating a 'balanced document'.

> We are not here to reach a conclusion on a piece of paper. We are here to build a model of agriculture that can support positive social change for people on the ground. We need strong and useful results that we can bring

> home to our constituencies around the world and support their struggles to build the model of food sovereignty. These issues are about our livelihood.
>
> (Latin American LVC member, CFS-rai negotiations, 20.05.2014)

In a self-assessment of LVC's first year of engagement in the CFS, the movement recognised that it is a great challenge for social movements to engage in the UN system where you can only take limited actions. This challenge was noted in an LVC evaluation after the first years of engaging in the CFS:

> For social movements, participating in the CFS is an immense challenge for which they are not necessarily prepared, due to in particular the lack of familiarity with the culture of negotiations and the tiny steps forward that lies at the heart of negotiations in a multilateral system.
>
> (LVC, 2012a: 14)

The following sections further explore some of the personal challenges faced by peasant delegates to the CFS, in particular regarding the mistrust vis-à-vis decision-makers.

b Mistrust

Convincing policy makers to support their proposals in CFS and building allies with governments is a central part of the lobbying strategy in a UN arena, where governments are the decision-makers. During CFS meetings, civil society activists actively seek to strengthen their positions by lobbying governments through side-events and informal meetings with country delegations, where positions are shared before official CFS sessions. Duncan found in an earlier study of the CSM, that alliance building with governments during the first years of the CFS 'proved to work in favour of the CSOs, especially when countries are supporting their statements' (2014: 161). While building alliances with governments is an efficient strategy for civil society groups to win support for their proposals in a UN intergovernmental context, it can be difficult for social movement delegates to accommodate the 'mindset of interest lobbying' (Ibid.: 149).

Not only are social movements often intellectually and emotionally tied to the positions of their organisations, their members also enter the UN arena with different political and cultural backgrounds and different personal experiences in facing political leaders. People and peasants who have been under the domination of local power elites or post-colonial regimes do not always feel comfortable building alliances with 'former enemies'. This is both a political and emotional issue. One member during the CFS-rai negotiations explained why he was not at ease with building alliances with governments:

> I do not trust them; they come to us and say they work for our interests, as peasants or farmers. But then they turn over to other government

> representatives to buy support from each other. Did you see how they just made a deal?[15]
>
> (South-Asian LVC member, first round of CFS-rai negotiations, Rome, 20.05.2014)

This mistrust was confirmed by another LVC 'newcomer' to the CFS, evaluating his first session in the CFS:

> When I am here at the CFS I am ask myself: what is happening behind the scenes? I don't feel we can trust them [governments] here. I am particular concerned that the private sectors go directly to the governments to negotiate with the decision-makers instead of discussing with us in the plenary. I don't like it when we talk so much with governments...can we really trust them?
>
> (European LVC member, 'newcomer' to the CFS, interview, Rome, 15.10.2013)

The experience with governments is often ambiguous for some social movement activists. On the one hand there is a strong awareness of the importance of the need of advocacy and lobby work vis-à-vis governments. On the other hand, social movement activists often show the 'uneasiness' of moving into the policy arenas dominated by states.

Whereas governments may present promising speeches about human rights when they speak in front of the world community and other governments in the UN, LVC delegates to the CFS often refer to the gap between discourse and practice. One LVC 'newcomer' to the CFS expressed his frustration during a CFS Annual Meeting:

> The same governments that talk about human rights one day here [in the CFS] sell the best land in their countries to private enterprises the next day. They all have hidden agendas. You must not forget that state and capital often work together in complex and non-transparent ways to serve one another. Here we sit face to face with our governments and we must use this and reveal what is really going on on the ground and that they are responsible to protect and support their citizens who give them their legitimacy.
>
> (European LVC member, 'newcomer' to the CFS, interview, Rome, 15.10.2013)

In the following section, I conclude with the argument that for agency to fully play out in the context of the CFS arena, participants must possess certain strategic skills and human capacities.

c 'Hold the line': the strategic capacity to act in the CFS

One main challenge for social actors to engage strategically in international negotiations is to acquire a range of personal skills or capacities. For civil society actors to engage in an effective way in the UN negotiations they need some degree of strategic intelligence of the unfolding power relations and strategic games of the policy process (Berry and Berry, 1999). This is particularly pertinent when negotiations speed up and the participants need to reach an agreement in a coherent way. LVC-technical support Nico Verhagen reflects on how moving strategically within an international UN arena requires a strong personal capacity from members:

> It is central for the movement that you have a good understanding of the policy processes. You strategically need to know how to follow that space, build it up, and have a good sense of what is at stake. You need human resources to do this type of work.
>
> (LVC technical staff support person, Skype interview, 30.05.2014)

Ideally, the individual must have the analytical capacity to understand the issues at stake and the cognitive skills to decipher how international negotiations play out. Acting strategically in a UN arena requires not only a solid understanding of the ongoing processes. It also means that the individual must be capable of analysing fast and be ready to move fast to shape the negotiations in the desirable direction. A Latin American member appointed to the CFS describes what this means to her in practice:

> When governments are coming to us, what does it mean? We need to analyse the context and see what concrete proposals mean. We constantly need to pre-empt and react quickly. Sometimes proposals on paper look good, but we must be on alert that it is not a Trojan horse for industrial farming. The devil is in the detail. These may be attempts to destabilise our strategy. We need to be alert.
>
> (Latin American LVC member, experienced CSM delegate, interview at CFS41 Annual Session, 15.10.2014)

The participants must have a clear understanding of the strategy and the space they move into in order to make decisions, define the terms of engagement and protect the agenda and articulation as a movement, so that decisions taken at a global level do not affect the profile of the movement negatively (Brem-Wilson, 2010).

Winning support for proposals in a UN setting does not only mean having a good lobby strategy ready. Training is also required to ensure that participants can carry out the high-level lobby strategies in practice and increase the personal capacity of actors to actively follow and influence the overall process during negotiations when negotiations advance quickly. Observations during CFS sessions and CSM meetings demonstrate that in practice 'full' participation ideally means that the social movement delegates participate in a myriad of formal and informal meetings to strategise, network, lobby, do internal assessments, and co-build to be prepared for CFS sessions. While civil society activists in the CSM group spend hours in preparatory meetings to strategise, they cannot always pre-empt how processes and negotiations will unfold.

> You have a strategy, and are prepared for that but you may need to react quickly, members go to another meeting and need to readapt quickly. Then you need to reflect on what is your strategy, what is your mandate and what is your responsibility. You need to hold your head cool.
>
> (Latin American LVC member, interview CFS41 Annual Session, Rome, 15.10.2014)

As one LVC member from the global South expressed it in a CFS session, 'holding the line' can be very confusing in a plethora of meetings.

> We are jumping around from one meeting to another. What is the idea behind [the] strategic moves we make? We need to have not only an understanding of the *how,* but also the *why* we are doing this in different spaces here. Whether this means in negotiations or deliberations in the plenary room, in Friends of the Chair meetings, in Friends of the Rapporteur meetings, or lobby meetings in the corridors.
>
> (Latin American LVC member, interview CFS41 Annual Session, Rome, 15.10.2014)

One of the LVC staff support persons who has accompanied LVC in some of the CFS negotiations in Rome, emphasised how this is, to a high degree, a learning process for newcomers to engage in the CFS: 'This is for most members a great challenge. It is natural that we are going to stumble along the way. This is how it is to be new in these processes. We will learn from this' (Skype interview, 30.05.2015).

While peasants do not always feel comfortable doing this type of work, interviews with LVC members engaged in the CFS show that the actors often express a strong will and sense of responsibility to engage in the CFS with the aspiration to change the world.

d Self-sacrifice and the will to change the world

Besides the challenging task of reading and drafting piles of technical UN documents, a number of peasant delegates to the CFS interviewed for this book also expressed the physical unease of engaging in UN work. For those peasants who have their main activity in the field, it can be very challenging to sit down in meeting rooms and participate in meetings for up to 12–15 hours per day, sometimes 5–10 days in a row.

> I am not used to or enjoy working with a computer and sit down all day. It is difficult to adjust to the conditions here as well as the monotonous way of working. I am used to doing meaningful work and going to sleep very exhausted and very tired in my body, but in my soul and my spirit I feel relieved because I do meaningful work and when I go to sleep I am happy for what I am doing. I am not a desktop person.
>
> (East European Member, LVC, Rome, 15.10.2013)

Another challenge is that participation in the CFS requires actors to travel to Rome at specific times of the year, following the holiday calendar of Western diplomats rather than the seasonal agricultural calendar of peasants. Peasant delegates often explained in my interviews with them how it remains a challenge to adjust to the Western linear calendar compared to the circular calendar of most peasants. Non-European constituencies – the vast majority of rural constituencies – need to undertake intercontinental travel and adapt to the time zones in order to engage in meetings. As expressed by an Indian farmer reacting to the questions about how he felt being at his first meeting of the CFS Annual Meeting in 2014:

> It is harvest time at home these days. I need to plan and talk to my family on the phone to be sure my tomatoes are harvested. I am a food producer. It is a sacrifice for me to be here. Not only for me, but also my family. I cannot be on my farm and do my work at home. I must pay someone to work on my farm instead of me. We are different from NGOs and other kinds of 'staff' who are paid for this work. For me as a farmer it is a cost to be away from my farm. It can sometimes look contradictory that I use so much time in Rome talking about agriculture and then leave my work on the farm behind. I constantly need to justify it to my colleagues that I am doing this kind of work.
>
> (South-Asian, LVC member, personal conversation, CFS Annual Session, Rome, 20.10.2013)

As this LVC member explains, engagement in the CFS is not only a personal challenge for him but also a burden for his family, as they must take on extra work when he is away from his farm. The member explains how his wife finds it difficult to understand why he is away to do politics for 10–15 days in Rome:

> I need to leave my farm and livelihood behind. I am not paid or compensated for my loss at home while I am away from my field. It is not only cost for me as a farmer to be here, it is also a cost for my family. It means that my wife and family members need to take on my work.
>
> (European LVC member, personal conversation, CFS Annual Session, Rome, 20.10.2013)

While using time and resources to engage in global policy work may in some aspects seem self-contradictory, that being in Rome may for some be a self-sacrifice, it is also a strong responsibility: 'You cannot feel equally comfortable in both worlds. However, we do this work here in the hope that our sacrifice and efforts will support our struggles on the ground and the lives of the generations to come.' Another LVC member explains that some degree of self-sacrifice is a part of the struggle.

> Being a social movement activist is the same all over the world: your commitment to the cause comes at a personal and financial cost. But the hard work needs to be done. Once we have decided to occupy a policy space we must be well organised and arrive to the negotiation table with well-deliberated contributions and solutions. We have a responsibility vis-à-vis other members to work for positive social change on the ground.
>
> (South Asian LVC member, personal conversation, CFS Annual Session, Rome, 16.10.2013)

While peasants do not always feel comfortable doing this type of work, LVC members often express the importance of the need to 'fill out the CFS space' that would otherwise be occupied by their adversaries. An LVC member expressed this during a breakfast meeting ahead of an Annual CFS session.

> We cannot let the space be filled by others. We have opened it, now we need to keep occupying it. I am aware that the CFS becomes more and more important and I know many NGOs would like to have my seat here ...We need to be here to protect what we have been struggling so hard to win. We need to be here, the struggle is bigger than ourselves. It is a struggle over which agricultural model and development path we wish to see in the future.
>
> (Observation, breakfast meeting with one LVC member, preparation meeting for the CFS Annual Session, Rome, 18.10.2013)

The different personal challenges and costs incurred by the actors can in some ways be outweighed by a strong sense of worthy self-sacrifice for the struggle that is bigger than the actors themselves.

> It is a struggle against the corporate model and the very existence of the land. It is a struggle for livelihood. People all over the world are facing fear at home in their everyday lives. Our land and resources are being grabbed every day over the world. In my country, oil and gas companies are taking over and are supported by the governments. This is why we not only have a reason but also a responsibility to take the struggle at this level.
>
> (East European LVC member, interview, CFS40, Rome 18.10.2013)

4 Conclusion

Peasant activists have different personal skills and arrive to UN meetings with different experiences and personal backgrounds. Those delegated traveling to Rome to represent their constituencies in international meetings do not feel equally comfortable engaging in a formal intergovernmental arena.

This chapter showed that, despite various barriers for agency to play out, social movements' engagement in an institutional setting does not automatically lead to a void of innovation and agency, as most social movement theories suggest (Tilly, 2004: 156; Tarrow, 1998; Blumer, 1969). For instance, the observations of rural activities participating in CFS negotiations from within the larger CSM group reveal how the autonomous civil society mechanism to a high degree functions as a laboratory for civil society participation in a UN Committee. In particular, for radical social movements engaging in a formal UN arena while seeking to resist the traps of institutionalization, the self-organising in the autonomous CSM is important to build a strategic engagement and a room for constant reflection and evaluation. The collective engagement and experiences from within this civil society space contribute to a heightened sense of agency, in particular among newcomers to the group.

Exploring how social movements' agency dynamics play out in the CFS via the CSM is, however, not to suggest that agency is free-floating,[16] neither is it to ignore the inherent powers structure and persistent barriers for participation. The UN logic (expressed in the dominant language, mode of communication, norms, code, etc.) remains prevalent in the CFS arena, and barriers for social movements' 'full' participation are still in place. In this regard, Brem-Wilson (2015)[17] documents how the 'burden of adjustment' for 'effective' participation is falling predominantly upon the shoulders of the participants themselves, i.e. their capacity to participate and adjust in accordance with the prevailing modes of communication and norms in the CFS.

However, the established UN structures and norm sets are also changeable. As Giddens argues in his 'theory of structuration': when actors are faced with social structures such as traditions, institutions, moral codes, and established ways of doing things, these may be changed when people start to ignore them, replace them, or reproduce them differently (Giddens, 1984). Consequently, future analyses of social movements' participation in the CFS should not only focus on the capacity of social movement participants and their organisations to work in accordance with the UN efficiency. This future pattern of engagement also depends on how the burden[18] of full engagement will be shared (Brem-Wilson, 2015: 19; see also McKeon, 2009a) and which kind of efficiency will prevail. How the actor–structure dynamics in CFS will develop in the future depends on the unfolding processes and is particularly pertinent for researchers wishing to explore patterns of exclusion/inclusion in transnational policy-making.

Notes

1 'Full participation' in the CFS processes excludes voting right.
2 As presented in the introduction of this study, Giddens defines agency as 'the capacity to act otherwise [...] to intervene in the world or to refrain from such intervention, with the effect of influencing a specific process or state of affairs' (1984: 14)
3 The main argument is presented on p. 4.
4 The difference between the World Bank led 'PRAI' and the CFS-led 'rai' (in lowercase letters) has caused confusion for members not following earlier negotiations.

5 English, French, Spanish, Russian, Arabic and Chinese. In comparison, during LVC's International Conference held in Jakarta 2013, translation was provided in 15 languages.
6 This observation raises questions for further empirical research to better 'bridge' between the different world spheres of social movements and UN arenas: how could this be addressed in future? Could a separate style guide be issued for interpreting food sovereignty movements? How would this work within the UN framework?
7 A 'cultural misunderstanding' can occur when words or gestures have different meanings in different cultures (Avruch and Black, 1993: 131–145).
8 This episode took place during the afternoon of the third day of the negations when it was clear that the negotiations would not be finalised within the planned timeframe of the CFS. The CFS Chair (not the working group Chair, who was leading the negotiations) went down to the CSM group and invited a delegation of the group out of the 'Green room' where the negotiations took place. In the corridor she thanked the CSM for their contributions, but kindly requested the group to select one spokesperson to intervene during the sessions and to remain focused on the text, in order to reach to a final document by the end of the working session. The technical support team to the CSM policy-working group kindly reminded the CFS Chair of the rights of the civil society group to self-organise their interventions. This observation illustrates the strong pressure that is exerted on civil society to adhere to UN efficiency ('be brief, stick to the point') in order to reach to a consensus document (Field observation, CFS-rai, 20.05.2014).
9 For instance, expressed by the FAO Director General in his message to the CSM ahead of the CFS-rai negotiations (Observation, CSM Annual Forum in 2014. 8.10.2013).
10 Communicated to me by a UN diplomat from a Latin American country at the UN headquarters, New York, 01.12.2012.
11 There is now a fairly set team working at the CSM on a regular basis with these interpreters having quite a high number of paid days per year (LVC/CSM interpreter, interview, Rome, 30.05.2014).
12 Communicated to me by a LVC/CSM interpreter, interview, Rome, 30.04.2013
13 Expression used by a CSM interpreter explaining that they agree to show a strong availability and flexibility when they work for the CSM (LVC/CSM interpreter, Skype Interview, 30.05.2014).
14 https://sites.google.com/site/interpretationspider/
15 By 'doing a deal' the peasant member refers to the strategy of governments buying support for their standpoints from the diplomatic logic of 'You do me a favour of supporting us, and I do you a favour next time'. This logic of supporting each other as a 'deal' was explained to me by a UN diplomat during a UN session in New York, December 2012.
16 Cf. Giddens (1984): the structure is transformed while actors are acting within it; it is the repetition of the acts of individuals that reproduces the structure.
17 Brem-Wilson (2015) presents a theoretical frame to address the issues related to 'arena adjustment' and barriers for producer constituencies seeking to convert their formal right to participate in the CFS into substantive participation.
18 For example the debate presented earlier in this chapter about who should pay for the translations to move closer to the ideal of 'full' participation of social movements delegates.

6 Reaching outwards and looking inwards

Influential social movement scholars directly relate the internationalisation of social movements with de-radicalisation on the one hand and a growing gap between an elite and 'the base' on the other (Tarrow, 1998: 134; Tilly, 2004). The expectation is that leaders of social movements will establish comfortable relations with authorities at the expense of the interests of the people they claim to represent.

> Professionalization leads to institutionalization, hence to declining innovation in social movements [...] social movement activists will sell out the interests of truly disadvantaged people, establish comfortable relations with authorities, rely increasingly on support from the rich and powerful, and/or become social movement bureaucrats, more interested in forwarding their own organizations and careers than the welfare of their supposed constituencies.
>
> (Tilly, 2004: 156)

The concern that institutionalisation will lead to a strong elite profile of activists is shared by a range of social movement scholars (Tarrow, 1998; 134; Tilly, 2004; Friedman, 1999; 396; Bellier, 2013; Meyer, 1993). For instance, in his article 'The Indigenous Struggles and the Discreet Charm of the Bourgeoisie', Friedman (1999) argues that delegates of the indigenous peoples' movements who participate in UN negotiations on the rights of indigenous people have become susceptible to forming an elite. The author describes how indigenous people and other groups engaged in the UN field are pulled by contradictory forces, in particular related to their hunt for funds, which has created a global elite of cosmopolitan indigenous leaders and the 'destruction of local accountability' as a result (Friedman, 1999: 406). Irene Bellier explains the dilemma related to the 'massive participation' of indigenous people in the international arena as follows:

> On the one hand, indigenous delegates in the UN are often perceived (outside the UN) as being absorbed by the machine (what Ferguson [1994]) calls an anti-politics machine), too much attracted to the

> international life and too loosely attached to indigenous localities to be considered 'true representatives.' On the other hand, they are efficient in forming an indigenous voice and in determining issues on the agenda of the international community as well as multiplying the recommendations adopted by the Permanent Forum.
>
> (Bellier, 2013: 193)

This chapter explores the practices and attitudes of peasant delegates to the CFS, and in particular, how these members seek to build models that 'connect' clearly with their constituencies. This is a path that differs from previous documented experiences and the expectations of a 'classical path' of internationalisation where institutionalisation is directly linked to social movements' elite formation, potential loss of contentiousness and innovation (Tilly, 2004; Tarrow, 1998).

This chapter illustrates some of the challenges for rural movements in building a model of internationalisation based on internal democracy and democratic leadership, while adhering to the movement's core principles, such as inclusiveness and consultation. A central element of the debate is whether peasant leaders engaging in an international context are representing the claims of their members. Based on the observations presented in the previous chapter, this chapter explores the challenges for LVC delegates appointed to the CFS, who are expected to translate between the global policy work and grassroots realities.

1 Internal democracy

a Accountability and representativity

As Chapter 2 showed, La Vía Campesina relies on a decentralised structure and derives its vitality and legitimacy from its member farmers' organisations at the local and national levels (Desmarais, 2007: 21). This section explores how building a model of action where members are selected by and are accountable to their members, requires an ongoing effort to address the issues of internal accountability and representativity. Several observers have noticed how one of the most difficult challenges faced by coalitions of highly heterogeneous associations is how to fully represent the diverse, and at times conflicting interests of their constituencies, and how to remain accountable for delivering the victories promised or for remaining true to the claims made (Borras, 2004: 24; Borras and Franco, 2009; Scholte, 2011; Desmarais, 2007). As expressed by Borras, Edelman and Kay in a study scrutinising the rise of transnational agrarian movements:

> Studying dynamics of interconnectivity – or the absence of it – between the international, national and local levels of contemporary agrarian movements, issues of representativity and accountability of these different

> levels tend to be sweepingly assumed rather than systematically problematised and empirically examined.
>
> (Borras et al., 2008: 11)

Although LVC has endeavoured to push out previous mediators, mainly NGOs and IFAP, this 'takeover' does not mean that they have completely solved the problems surrounding the lack of full and real representation of local villages and national groups at the global level (Borras, 2008: 13–18; Borras and Franco, 2009).

Today, social movement delegates move into different venues and arenas at a global scale. This brings about new challenges that they need to face up to. In the context of the CFS, data from fieldwork show that the issues of representativity and accountability are addressed by LVC leaders themselves on a regular basis. At one of the daily evaluation meetings during the CFS-rai negotiations, one LVC member posed the question: 'How do we effectively make sure that we represent the perspectives, experiences, aspirations, values, visions, and ideas of our members, while we engage in transnational global policy work?' (African LVC member, CSM evaluation meeting CFS-rai negotiations, 16.08.2014).

While the issue of representativity may be in the heads of social movement leaders during global meetings, in practice the nuances of 'who is representing whom' are often left out in the actor's discourse. A civil society activist engaged in the design of the CSM explained this to me. The very nature of the CFS arena remains a political battlefield with strong imbalances of power. Moreover, CSM members have to face many complex issues; they are under pressure because of the many overlapping responsibilities they bear as they move into a complex global governance terrain. As an LVC member clarified, when I interviewed him after a CSM evaluation session in 2013:

> We need to be more careful when reflecting on whom we claim to represent and when. It is a challenge to train more members to understand how this complex infrastructure works. We work across different networks and different constituencies around the world.
>
> (LVC member, interview, Rome, 16.10.2013)

It is crucial for LVC members to address issues of internal democracy, including representativity and accountability, as they seek to build a 'tight but flat organisation' (cf. Chapter 2). This leads us to the debate of how transparency, consultation, accountability, and self-evaluation become key components of 'democratic accountability' when LVC leaders engage in the CFS. One LVC member explained it in the following way:

> Accountability and authentic governance mean that social leaders are responsive to their membership, primarily accountable to that membership, and secondarily accountable to whatever external processes or institutional

> partners they are working with. For us, authentic governance requires members to be active participants. It means that the system of governance must facilitate active, collective dialogues and decision-making.
>
> (Personal conversation, English-speaking LVC member, Rome, 13.10.2013)

The following sections show how LVC is aiming to build a democratic leadership model that is based on a high level of 'reflexivity agency' (e.g. that actors have the capacity to reflect on their own action; see Giddens, 1984). Before I present how LVC seeks to build a model that is based on a high level of dialogue, internal critique, and self-assessments, I discuss how 'prefigurative activism' (consistency between one's values and one's practices) is a central dimension of activism in many movements (Pleyers, 2010: 142: see also Epstein, 1991; McDonald, 2006).

b The challenges of internal democratic leadership

Despite the often strongly declared values of internal democracy and prefigurative activism, Pleyers reveals that social activists do not always meet these ideals in their practices (2010: 142). For instance, studying the pattern of activists engaged in the *Alter-globalisation movement* in the period from 2000 to 2010, Pleyers (2010) demonstrates how the World Social Forum (WSF), in particular during its first three versions, was far from the open, diverse, and democratic space it claimed to be. In particular, the author revealed how the WSF-International Council (WSF-IC) lacked principles of representativeness or participatory democracy and was centralising power in the hands of a group of intellectuals (2010: 151). His findings show that the organisation around the WSF-IC in this period was legitimised by leaders who saw the need for better 'management' of communication and coordination of international gatherings, and the 'urgency to find responses to neo-liberal globalization' with the result that concerns for internal democracy became secondary (Pleyers, 2010: 142–53;[1] see also Santos, 2012: 90–95). While the WSF later responded to this critique[2] (Pleyers, 2012; Caruso, 2013), fieldwork within the food sovereignty movement shows that some activists, including some LVC leaders, still question the legitimacy of the WSF (Vieira, 2011: 217). LVC leaders interviewed for this study are themselves highly conscious of the risks of 'classical' leadership formation (Tilly 2004: 152) where a handful of strategic leaders monopolise the global debate and de-connect with the base.

> Professional activists tend to occupy the role of intermediaries between citizens and institutions, filtering the discourses on both sides, defining the strategies of mobilisations and reacting to public policies. These newer, lighter externalised movements no longer have a permanent cadre of grassroots activists.
>
> (Tarrow, 1998: 134)

The tendency to elitism and lack of transparency when social movements internationalise their struggles has been presented as one of the main reasons for the weakening of other social movements (Vieira, 2011; Hyman, 2002; Waterman, 2014). Such historical accounts of elite characteristics and undemocratic leadership[3] help to explain why LVC members zealously care for internal democracy while building its model of internationalisation.

LVC leaders engaged in the CFS and interviewed for this work often explain how 'connecting with their base' is a condition for their engagement in the CFS. An African LVC member and 'newcomer' to the CFS clearly expressed this an interview during the CFS41 annual session in Rome.

> It only makes sense for me to be here, because it is anchored in my daily work mobilising against the massive land and resource grabbing we experience in Mozambique. Gaining support to our member base must be our first concern. We need to negotiate tools for people to take back home to the national organisations that can support the struggle there. We must find ways to build bridges.
>
> (African LVC member, interview, CFS41 Annual Session Rome, 15.10.2013)

Several LVC delegates to the CFS often explained how the LVC movement is built around a 'collective voice, not an individualist one'.

> I am not here on my own. I cannot act here from an individualist point of view. I am entrusted to be here to fight for the collective movement struggle. You represent a message from your peers and group; they are with you generating your team message. We must always remind each other about that during the meetings here.
>
> (European LVC member, interview, CFS41 Annual Session, Rome, 16.10.2013)

Observations of and interviews with LVC peasant leaders who are delegates to the CFS reveal that the responsibility that comes with being an 'entrusted leader' is often on their mind when they engage in global policy work. A 'newcomer' to the CFS-rai negotiations in 2014 acknowledged this in the following way when I asked him how he felt about being a spokesperson for the LVC at the global level for the first time:

> I find it difficult to sleep well ahead of a plenary session. I often wake up during the night because I am afraid that I will forget something. This is not a work or responsibility of my own, but the work and responsibility of thousands and millions.
>
> (Asian LVC member, CFS-rai negotiations, interview, 13.08.2014)

The examples presented above illustrate different examples of how LVC members who have been appointed as delegates to the CFS often take their duty and responsibility very seriously. However, being a spokesperson to engage in global policy process can be a complex, demanding and challenging human task to complete.

One of the main surprises throughout this research has been the degree to which LVC leaders are open to talk about and reflect on their own roles as leaders in the movement. For instance, at several points during this research, LVC members contacted me on Skype to discuss their role and the experiences that stemmed from their engagement in global policy work. This openness and willingness to engage in self-assessment are signs of reflexive[4] and democratic leadership as opposed to the hierarchal leadership that is predicted by theories on the professionalisation of social movement leaders (Tilly, 2004; Tarrow, 1998).

For LVC, building a democratic model of leadership is coherent with the movement's strongly held principles of participation and horizontalisation. Building this model of internationalisation based on a low degree of institutionalisation (cf. Table 2.1) is also important to the internal construction of the movement. Paul Nicolson, a Basque farmer and one of the founding members of LVC, expressed this in an interview: 'In our movement we cannot have leadership in the traditional sense. We need to build our leadership around another type of leadership. A model where leadership means leading with peers' (Paul Nicholson, Skype interview, 30.10.2014).

One of the LVC 'newcomers' to the CFS explained at a CSM meeting how debate on their own role as movement leaders can actually help them and their movement to become stronger.

> It challenges you to look beyond formulating quick technical solutions, and instead asks you to examine yourself. This is important for your own commitments and the loyalties you have vis-à-vis your constituency. This is complex but self-assessment is a good way for us to start to move ahead and makes us all stronger.
>
> (LVC member, CSM meeting, Rome, 10.10.2014)

This awareness of having to adhere to strong values of self-assessment and of the responsibility of finding ways to 'connect' with the broader movement contrasts with the predictions made by dominant social movement literature, which argues that leaders are expected to become cut off from their grassroots constituencies (Tarrow, 1998; Tilly, 2004). The strong willingness of rural leaders to discuss their own roles as leaders plays an important role in building another model of leadership and internationalisation.

The following section explores some of the practical challenges for LVC delegates to engage in a UN committee, in particular the challenge of translating back and forth between global advocacy work and local struggles.

2 Linking global policy arenas and peoples' struggles

a The role of the translators

Chapter 2 of this book discussed how civil society participation (read: NGOs) in global governance spaces in the 1990s was largely dominated by NGOs with the required resources (e.g. Burt, 2005; Mosse and Lewis, 2006). This led to a trend that marginalised peasants and their ability to articulate their own needs, interest and demands at the global level (Desmarais, 2007: 9; McKeon, 2009a). As a people's organisation with a strong focus on representativity and accountability the very struggle of LVC has been to break with the dominant NGO model and shift towards a model of more direct mediation. Yet, contributing to transnational policy processes – and bringing back these processes to the ground – remain highly complex. This can be seen, for instance, in the reflections made by an LVC member engaged in global policy processes in Rome:

> How can I better get our constituencies with us here? How can we bring the issues considered most important at the grassroots level up to the CFS for inclusion in policy? How can we ensure these are not individual demands that we are discussing here, but that they are shared grassroots' struggles? It means that we must get a mandate from our groups so we can deliver these messages. It is a challenge, as we need to deliver rapid responses in international meetings to get our message across.
>
> (European LVC member, interview conducted after a CSM wrap-up meeting, 08.08.2014).

Bringing the local demands to a global policy arena like the CFS is only part of the work of the 'translators', seeking to build the local–global infrastructure between grassroots struggles and sites of global governance. After participating at the global level, leaders of peoples' organisations are expected to report back to their constituencies. A LVC member expresses during a civil society evaluation meeting how this is not an easy task:

> How do I really share what is happening and what is going on in global policy processes that are so complex, so that the grassroots members see the relevance? How do we communicate this? How do we translate the messages into an understandable language so the documents make sense to the people they intend to serve? How do we share what we are doing here? How do we provide feedback?
>
> (European LVC member, CSM wrap-up meeting, Rome, 08.08.2014)

Peasant leaders delegated to the CFS often state that global engagement can only be of true value if it is based on genuine participatory outreach to, involvement with, and feedback from the local level.

> Social change does not start from here. The train keeps running outside. We must find ways to bring Rome back home. We need to keep building these links better. It is a long process between the local struggles and global work. We are in a process of experimenting, learning from each other – and in a process of finding out how we can build the links.
>
> (European LVC member, CFS40, Annual Session, 18.10.2014)

By reflecting on the political values of the movement, LVC leaders delegated to the CFS strive to have an expressive and responsive leadership:

> Not only must grassroots see the relevance of what we do in Rome. We must go one step further and not only make sure that we achieve high level victories relevant to the grassroots, but also ensure that we are ultimately accountable to grassroots members.
>
> (Latin American member of LVC, CSM meeting, Rome 30.05.2014)

As we will see in the following section, one way to support this linkage process is to develop effective channels of communication within the movement.

b The role of information and communication technologies

In Charles Tilly's account of social movements' internationalisation, factors such as uneven access to information and communication channels can lead to an increased gap between social movement leaders and the member base of the movement (Tilly, 2004: 155). While access remains uneven, social movement activists are increasingly active users of information and communication technologies and skilled in using these channels to build their relationships (Karatzogianni, 2006: 55). In particular, during the last two decades, more activist groups are taking advantage of networks to engage simultaneously in multiple sites (Castells, 2000: 502). LVC is an example of a global movement making active use of information and communication technologies in building its global network and model of internationalisation. While the use of communication tools does not necessarily mean that the challenges of participation, representation, and accountability can be overcome, however, it can help to build the internal infrastructure in movements by facilitating connections at the transnational level (Castells, 2000; Sassen, 2007).

Most social movement representatives, but not all, enter the CFS with a laptop. Access to a computer and being online during meetings helps group members of the CSM to rapidly share documents and connect with other members during sessions. Fieldwork observations also show that some of the LVC members seeking to reach out to constituency members back home are actively using social media platforms such as Facebook and Twitter. For instance, one LVC delegate to the CFS sitting next to me during a CFS session explained that he was writing Twitter and Facebook updates, and that he used a blog to regularly upload short articles with photos in an effort to increase transparency and give others the opportunity to follow the UN processes from abroad. The LVC member, participating in the CFS annual session for the second time, explained that he found this communication important, as it was a way to 'report back and inform other members of what they are doing' and 'help others understand what is actually happening in the UN'. The LVC member elaborated on this and stated:

> I think over time, without good communications processes, the rift widens between the insiders and the grassroots they represent. We must continue to work on all aspects of what we understand by good governance, when we engage in this kind of space. I think we are learning a lot from each other here and must ask ourselves how we can do this better.
>
> (African LVC member, interview, Rome, 13.10.2014)

The use of the CSM as a space for intra-learning confirms that this global coordination mechanism, to a high degree, functions as a laboratory for civil society activists seeking to find their feet within the UN arena and disseminate what is happening to their constituencies 'back home'. When representatives of people's constituencies actively seek to 'connect' the UN and grassroots organisations, they contribute to the production of a re-connection between previously unconnected or weakly connected sites. This can be seen as an example of peasant activists seeking to build the global–local infrastructure and the 'scale' between the local and global (Bringel, 2015; Pleyers, 2015: 110).

However, the use of new technologies also has its limits in building the internal infrastructure within a peasant movement. Online communication does not reach out to all 'layers' of the movement.[5] This explains why face-to-face meetings remain an important priority at the local, national, regional and global level and the importance of building spaces to debate these issues in the broader movement. Bringing the CFS engagement onto the agenda of the movement's 6th International Conference is an example of how LVC is seeking to broaden the debate about UN engagement to the wider movement (Fields notes, LVC's 6th International Conference, Jakarta, 15 June 2014).

Yet again, while facilitating face-to-face meetings supports a growing articulation between the global level and the base, it is, however, simply not practically possible to convene a personal meeting every time an act of communication is required. Such practical limitations of inclusivity corroborate that some degree of delegation is needed[6] and explains why LVC organises as a people's organisation where members are delegated to take on responsibilities and 'entrusted leadership'.[7]

c The challenge to 'bring Rome home'

As Desmarais notes, as a global movement LVC largely derives its vitality and legitimacy from members of peasant organisations at the local and national level (2007: 21). Data from fieldwork show that LVC members, who have been appointed as delegates to the CFS, often express the importance of providing feedback in both directions, between global policy processes and peoples' struggle on the ground. Yet, despite a strong consciousness of the importance of dissemination registered in the section above, LVC leaders continue to face the ongoing challenge to translate technical UN texts into 'peoples' language' and translate between different political cultures (Santos, 2012).[8] One Latin American LVC member delegated to the CFS explained to me what she experienced after returning home after a CFS session. On a Skype call a week after a CFS-rai session in Rome, the LVC member started our conversation by explaining that she could only spend half an hour on the interview as she was expected to do her regular briefing from the CFS session to her own organisation immediately after our talk. She went on to explain the challenges of reporting back to members of her national organisation after each session in Rome.

> We need to translate what is very complex into movement language so that peasants in our communities see the relevance of global policy work. It can sometimes be really difficult to get this to the ground level and make the national movements understand what is happening here in Rome.
>
> (Latin American LVC member, Skype interview, 30.10.2014)

Another LVC member delegated to the CFS added the challenges related to the dissemination of technical UN work 'back home'. She stressed the importance of not giving the impression of being superior to other members when returning after global meetings with a large package of acquired knowledge, international experiences and sometimes personal relations to policy-makers and other elite actors.

> I must do this feedback without patronising anyone. Even for those working here in Rome it is very complex. The language is very high-level. So, imagine how difficult it is for us to go back with this. This work is very high-policy. The question is: how much can we really disseminate?
>
> (South Asian, LVC member, experienced CSM delegate, Rome, 16.10.2014)

These challenges reflect the dynamics of the ongoing challenges related to dissemination and 'translation' (Santos, 2012), as previously identified in Chapter 5, where we examined how the issues of language and translation in different aspects were still the main barriers to 'full' participation for social movement activists representing food producer constituencies in the CFS. The experiences expressed by LVC delegates above corroborate that they must navigate the different levels and address existing structural barriers in order to build global–local linkages.

The following section further presents how LVC faces a number of practical limitations as it seeks to build its model of internationalisation by increasingly engaging in multi-scale activities.

3 Between political imagination and practical limitations

a Short versus long-term victories

A central goal of social movements' engagement in lobbying and advocacy work is that these can be *means* through which social movements build support for peoples' struggles on the ground. Yet, social movement delegates appointed by their constituencies to engage in UN processes are frequently faced with the challenge that their engagement does not always bring tangible results that they can 'share' with their membership base.

Rural activists employing a great deal of their time in global meetings explain the tension that they need to justify to the constituency members when they use their time on global work far away from people's struggle. 'We need good results to bring home with us' is a sentence often heard when LVC engages in CFS sessions. Concrete decisions are important in order to gain political credibility in the eyes of the social movements (LVC, 2013). This pressure to 'bring home good results' is also registered in several interviews conducted with LVC members delegated to the CFS. Several LVC members explain that their organisations continuously question why they use a great part of their time on global policy work at the UN far away from the ground. They ask:

> What do we get out of it? It can be difficult to convince people back home that the role of a peasant farmer in high-level policy at the UN is crucial. The policy work in the UN is not binding and it can be difficult to defend spending so much time on this. The outcome is not immediately seen.
>
> (East European LVC member, interview CFS Annual Session 18.10.2014)

In the context of LVC's engagement in the CFS, how the collective messages of 'victories' are delivered is crucial, as these play an important role in encouraging the shared vision of the movement (Fox, 2003). Paul Nicholson, known as one of the strategic leaders of LVC, shares his reflections on LVC's complicated engagement in long-term struggles:

> It remains an ongoing challenge to social movements to learn how to win, but it is equally important to learn how to lose. We must know when to win and when to lose. We must learn to distinguish between short wins and the very long-term wins.
>
> (Paul Nicholson, Skype interview, 30.10.2014)

However, the tension continues to reign, as peasants engage in the global sphere. For instance, when LVC representatives travel to international UN meetings they often bring with them strong testimonies and powerful messages based on their personal experiences. When peasant leaders return home, they come back with a pile of technical policy documents. A Latin American LVC member summed it up in the following way:

> Members from my organisation start to ask why I spend so much time going to Rome. What do the time and resources spent in a global policy arena bring to the struggles on the ground? Is it really worth all the effort and resources, they ask: Does this support our struggles here?
>
> (Latin American LVC member, interview, CFS40, Rome, 14.10.2014).

The pressure 'to show results' described by LVC member delegates at the global institutional level is also registered in the fieldwork conducted with LVC members at the national level. For instance, Delatorre Alvaro, member and technical supporter to the Landless Workers' Movement (MST) in Brazil, depicted the tensions building up during his over 20 years of experience working with the MST. The MST veteran explained some of the challenges of peasant leaders delegated to negotiate with government authorities: 'Social movements need successes. This is how social movements work. Leaders need to show that they bring about results to the movement. Otherwise you easily lose your credibility' (Delatorre Alvaro, interview, Santa Maria, Rio Grande de Sul, Brazil, 16.03.2014).

This tension was equally described to me in an interview with Flavia Braga Vieira, Brazilian professor and activist scholar with 20 years of experience working within the network of the Brazilian members of LVC. She affirms that those members participating in international meetings are to some degree 'controlled' by other members. Vieira is referring here to an example of working with the Movement of People Affected by Dams (MAB) in Brazil.

> Here in Brazil, members joke about those who attend international meetings too much. For instance, they are openly joking about those members who 'always have their suitcase ready to go to international meetings'. These are not just jokes. They function as effective moral mechanisms inside the movements. They do not set up rules or laws that say that member can only go to, for instance, maxim four meetings per year. There is a common understanding that there is a limit for how much absence people are accepted to be away from the local struggles. These mechanisms are strongly felt by anybody working within the movement.
>
> (Flavia Braga Vieira, interview, Rio de Janeiro, Brazil, 18.07.2014)

This testimony from Brazil reveals how members on the ground are conscious about the opportunity cost (the cost of lost opportunity involved when someone is absent from the national work to attend international meetings) of doing the international work. This means that those that go to the international meetings risk creating tension if they spend too much time at that level – and particularly if they return home 'empty-handed'. One LVC technical support member reflected on this dilemma in the following way: 'Even if nothing bad happens, if you lose too much time and this time is taken away from the struggle on the ground you may lose credibility vis-à-vis your constituency' (Skype interview, 30.01.2014).

Paradoxically, the great expectation of leaders to fulfil a role of political leadership within the CSM group (and not leave this space for NGOs) means a constant risk of a process towards 'professionalisation' where social movement leaders end up spending considerable time and energy on engaging in these meetings at the expense of time on the ground. If movement leaders spend too much time in global policy arenas, it implies that they run the risk of increasingly creating a hierarchy of those who are delegated to global work and those who are not.

b How much time to what type of work?

Engagement in the CFS means that LVC and its member organisations must carefully balance the (financial and human) resources that can be allocated to transnational policy work. As LVC states in a policy document after the first two years of the movement's engagement in the CFS:

> Social movements' engagement at this level requires continuous preparation as well as technical and communicative support to decode languages, carry out the work of drafting standpoints, deal with media, and other CFS related tasks.
>
> It means in practice that there is a constant risk that this work associated with the involvement in the CFS will divert time, energy and resources away from the movement's mobilisation and struggles on the ground.
>
> (LVC, 2012a: 12)

Consequently, members of LVC must constantly do the balancing act of calculating how economic and human resources can be allocated between time spent on mobilisation, confrontational positioning and building up alternatives on the ground. How should that time be allocated? This was one of the topics debated in a thematic meeting during LVC's International Jakarta Conference. Here, one of the LVC members emphasised that mobilisation remains the movement's main strategy referring to a thumb rule of '99 per cent to mobilisation versus 1 per cent to institutional work'.

> We must always remind ourselves that our main strategy is mobilisation. Most of the time we are struggling for the radical change on the ground. We retain our own strategy. But if just one per cent of the time we spend in this arena can support our struggles on the ground, even just a few people on the ground, we have a reason and a responsibility to be there [in the CFS]. If we are not in this space we know who will win, the corporate sector and we will see more industrial agriculture around the world.
>
> (Field observation, group meeting on 'governance', LVC's 6th International Conference, Jakarta, 15.06.2013)

While LVC leaders clearly state their various priorities in official discourses, this allocation of time is in practice far from clear-cut. For instance, LVC members of the CSM Coordination Committee interviewed for this work frequently state how they often end up spending much more time on the process than originally planned. Apart from the activities undertaken by the CSM Working Groups, there are groups that work all-year round via multilingual email exchanges and Skype meetings to connect civil society members across continents. Most LVC members have numerous responsibilities in different Working Groups and committees. Those participating in meetings related to the CFS are expected to be well prepared and ready to take on political leadership, participate in meetings and subsequently provide feedback and proposals on draft versions to the broader group. Time management between what is used on global policy work and what is used on struggles 'back home' is still a balancing act for social movement delegates. A South Asian farmer representing LVC in the CFS explained this paradox in an interview during a CFS session: 'When I spend time away from home, I constantly risk losing face on the ground. I constantly need to balance my time and engagement – if I use too much time at one level, I will lose credibility at the other' (South Asian, LVC member, interview, Rome, 16.10.2014).

This constant pressure to balance the limited time they have for lobby and advocacy work with their international obligations was clearly stated by an LVC member after the CFS-rai negotiations in 2014. During an evaluation meeting, CSM members were actively strategising about the 'next steps' that the group should take. One LVC member interrupted the meeting at one point with the following message:

> As social movements we cannot invest more time in the microphone. We are not here to become NGOs. We are part of the world here but social movements should remain movements. We need to focus – we have nine working streams in the CFS this year. If we want to be effective here this is too much. While we are enthusiastic when we meet here, we cannot open up too much. This requires a lot of work here. If we start to be here 365 days a year, we are no longer social movements. We must keep our souls, our way of seeing things and we must remain social movements. Now it is time to for us to go home and consult and evaluate the whole process.
>
> (European LVC member, observation, CSM internal evaluation, Rome 10.10.2014)

As the LVC member expresses here, it is important that the intense lobby and advocacy work will not mean that social movements fall into the trap of becoming institutionalized. There is a constant fear the leaders will 'start to act like governments or professional NGOs'. This concern was, for instance, articulated by an LVC 'newcomer' when he for the first time arrived at CSM and witnessed how the advocacy and lobby work in the UN take place lace in the corridors. During a CSM meeting, he expressed the following in an upset and frustrated tone: 'Why are you sitting with them [governments] all the time? You start working like them, you will start being like them, we should not become governments, we are different' (Observations, CSM evaluation meeting, Rome, 10.10.2013).

The paradox is that those delegated to the CFS arena, as representatives of rural constituencies, risk their relationship to their wider membership if they engage to much in global policy work. In other words, by performing the role of representation, they potentially become *less* representative. Such tensions affirm the importance of social leaders' ability to reflect and constant willingness to debate on how leaders can achieve this balancing act of simultaneously taking the lead in international negotiations and remaining authentically 'rooted' grassroots leaders.

c Between personal commitment and limited financial and human resources

Engaging in UN meetings and translating between UN work and grassroots struggles is a complex task for most peasant delegates. On top of their responsibilities, vis-à-vis their social constituencies to 'bring good results back home' and disseminate complex information, participants in the CFS are under constant pressure 'to get their work done' in Rome. An external evaluation of the CSM conducted in 2014, five years after the CFS reform, refers to the risk of 'overstretching' those representatives delegated to the CFS and states:

> Besides fulfilling their political mandate, those engaged in the CC [Coordination Committee] work of the CSM are expected to contribute to the internal development of the CSM and serve in a facilitating and organising role to connect their constituencies/sub-regions. Social movements are faced with the challenge of limited capacity, time and the difficulty of 'reaching out' to affected communities, including those with limited or no Internet access.
>
> (Mulvany and Schiavoni, 2014)

LVC members representing rural constituencies in the CFS do this work secondarily or thirdly on the top their work as food producers and responsibilities as leaders at national and/or regional level. LVC leaders engaged in the global work often emphasise that this kind of work is for the main part voluntary and on the top of other responsibilities that they have as national leaders. Members of the CSM Coordination Committee explain in interviews that from the outset they were not fully aware of how much time and energy it takes to fulfil a role of political leadership in the UN processes:

> It goes fast here. You need to get your work done. This means you need to prepare, to read piles of documents, to strategise, to participate, to do your lobby work. While doing all this you think of how to translate this back to your base afterwards so that it makes sense for those who are not here.
>
> (LVC member, Rome, 18.10.2014)

Dedicated activists often embody the tension between strong political commitment and limited financial and human resources. While peasant delegates to global work may show strong personal and political commitment to take on work tasks and responsibilities, they are faced with personal and organisational limitations to take on additional work. Judith Hitchman, who works for URGENCI (International Network of Community Supported Agriculture) and has 40 years of experience as an interpreter and translator within the international network of the World Social Forum, phrased it as follows: 'The single biggest challenge facing most social movements today is that they are all under-resourced and over-committed' (Hitchman, 2015: 10). In an interview, Hitchman, who currently holds the mandate of Consumer Constituency representative within the Civil Society Mechanism, elaborated on this tension:

> There is an excessive concentration of work and responsibilities on too few shoulders on a voluntary basis. Also, people who are insufficiently briefed or not fully aware of the contents and issues at stake are sent to meetings. In the first case there are several risks: too much power becomes concentrated and the risk of genuine horizontality is high, and in the latter the real effective impact of the social movements' input can be diluted.
>
> (Hitchman, interview, 03.07.2015)

This leads us to another dilemma, namely the challenges related to rotation of participation.

d The dilemma of rotation

If LVC wishes to take inclusive political leadership at the global policy level and 'fill' the CFS space with peasants' demands and voices, the movement must continue to send new leaders that can represent small-scale food producers and participate in the ongoing policy struggle at this level. Values such as transparency, democracy and inclusive leaders are often emphasised, along with the importance of rotation in participation. This aspect stuck out when I interviewed an LVC member in the process of exchanging seeds with another member. This very active leader whom I had observed, surprised me by telling me that he would neither return to the CFS process, nor participate in the international meeting: 'It is time for a new generation and it is time for me to go home and plant these seeds', he stated while showing me the seeds he had in his hands. This short episode outside the official UN negotiations rooms revealed not only the peasants' connections to seeds and the practice of producing many types of crops; it also showed the awareness of the reasons for the internal rotation among members of the movement.

> We need to sharpen our ability to influence international negotiations, emphasising the construction and defence of values that we want to see in

> the world. We need to mobilise better and in particularly young people to take on this kind of political dialogue.
>
> (European LVC member, CSF40, Rome, 10.10.2013)

For LVC, rotation of participation means continuing to mobilise new members to engage in global activism. Yet in practice, rotation is a constant challenge for the peasant movement: while frames, campaigns, agendas and interlocution become more global, a relatively small number of the movement travel to global meetings. Rotation is a classic dilemma for social movements. While a high degree of rotation increases democratic leadership for social movements, it also hinders some of the effectiveness, e.g. the need for some routine, continuity and historical memory (Pleyers, 2012: 180 n.11). A Latin American LVC member, engaged in the CFS since the reform process in 2009, expressed the need for some degree of continuity to work efficiently in global meetings.

> We always seek to include newcomers, but you can imagine how this is a challenge. You often feel you start all over again, when we start meetings by explaining the process to new members and what we have already gone through as a group. I know this is something that is important to do, but this also takes time, and we are under pressure. We need to get our work done.
>
> (Latin American LVC member, CFS session, Rome, 14.10.2013)

LVC members appointed to be delegates to the CFS hold high values of inclusion. This has been observed on several occasions during this research. For instance, during the CFS40 session in Rome I interviewed a young delegate from South Asia' he explained how he followed a more experienced LVC leader from the region during his first meeting, in which he participated as an observer. He admitted that he found this first experience instructive and described it as 'a comfortable way to learn to engage with the UN' (South Asian Youth Observer to the CSM, conversation, CFS41 Annual Session 15.10.2014).

During this course of research, I witnessed several instances of social activists choosing to prioritise inclusiveness over 'UN effectiveness' during several CSM meetings. For instance, CSM newcomers are often delegated to be spokespersons on behalf of the group, however, they are well aware of the fact that they are generally more nervous and may make their interventions in the CFS less convincing (according to the UN norm to speak clearly and briefly) than more experienced members within the civil society group. Such a working culture that focuses on inclusion, process and training requires, at times, a great deal of patience and tolerance from other members in the civil society group, perhaps particularly from NGO representatives, who are used to dominating global meetings (McKeon, 2009a: 176–177).

Even though tensions remain within the CSM, fieldwork observations of members in the CFS show that they regularly remind each other to nurture this culture of participation and inclusion that they wish to build. For instance, during the CFS-rai negotiations I observed how a newcomer to the CSM group apologised for forgetting to present a point during his intervention and how disappointed he was with himself. After the official session, the group united in the corridors of the FAO building to evaluate the CFS session, as it often does. At this point, a more experienced LVC member reiterated how important it was for CSM spokespersons to recognise that they are bound to sometimes make mistakes: 'This is how we move ahead. We were all new in the beginning. It is by learning from our mistakes that we all become stronger' (European LVC member, Rome, observation, Rome, 20.10.2014).

The training of new leaders, in particular women and younger people, to represent the movement at a national or local level, and perhaps in the longer run, at an international level, is an important objective of the movement. The need for more training was clearly stated in the report coming out of La Vía Campesina's 4th International Conference in Jakarta in 2014: 'There is the need to train more leaders to participate in international level politics without compromising the struggles at the local level. Women and youth are still to be fully integrated in all activities' (LVC, 2014a: 17).

As presented in Chapter 4, the very nature of UN work can be rather cumbersome, in particular for youth activists who tend to be attracted by the more expressive aspects of the struggle (Pleyers, 2010: 78). This explains why LVC is faced with a challenge to motivate the younger generation of LVC leaders to engage in formal policy work. As one youth LVC member participating in the CFS session shared with me: 'I am here trying to understand how this can make sense for our generation. UN work is just so anti-youth' (European youth member of LVC, CFS-rai negotiations, Rome, 19.05.2013).

The UN itself seems to face the challenge of how to make its work more 'youth-friendly' and more attractive for groups, who contribute to the legitimacy of the UN. One of the LVC technical staff pinpoints some of LVC's own hurdles, such as its goals for gender balance and increased participation of youth in global meetings, but he also reiterates LVC's determination to rise to these challenges:

> At this level we have many difficult issues. But it is not because these issues are difficult that we do not seek to resolve them. We know these are difficult challenges but they should remain at the top of the agenda. If you keep thinking about them and trying to resolve them, it helps you as a movement to be more flexible and dynamic and to avoid bureaucratisation. Because as soon as you give up on these kinds of issues, then you sit back and fall into the other easier tracks of organising with the people that speak English and have experience, as that is most effective (…) These issues are not technical problems to resolve. These are political questions. It is about what role we want to play in this, how we are limited and how we should organise ourselves to overcome it. It is a question of keeping the movement on track and avoiding falling into a much easier bureaucratic system.

> Of course this requires a lot of energy trying to solve it and some people are more likely to put in this kind of energy than others.
>
> (LVC technical staff support person, Skype interview, 01.07.2014)

Building a model of internationalisation that reflects the movement's values and principles means that the movement must persistently address how actions are undertaken in internal practices. For instance, not all peasant organisations fulfil their internal goals; not all have been successful in training new members to fulfil the declared goal of gender equality.[9] Such challenges are internally discussed on a regular basis. However, from the LVC's secretariat side, one of the technical staff support persons himself underlined that the technical support team can only advise; they cannot sanction members who do not respect the declared goals.

> We cannot decide what members do. In LVC the relation goes the other way around: the national members, who decide over the movement, control the movement. For instance, we cannot say that LVC does not pay your ticket if you do not report it when you come home. We can only advise them about what we find is for the benefit of the broader movement; meaning that the national organisation should make sure that women and youth are better represented at the global level and at meetings. LVC does simply not work like that…It is a culture of a movement that you try to build to up as a movement but internally and in interaction with others, such as allies and institutions.
>
> (LVC technical staff support person, Skype interview, 01.07.2014)

One way that LVC goes about building its structure is by providing an internal space for dialogue, ensuring that the values and political culture of the movement are ever-present. This seems to be particularly relevant during those periods when LVC engages in complex policy processes at the global policy level. Such internal spaces are crucial given the constant need for strategizing, evaluation and self-assessment on whether the movement is achieving its target model of internationalisation (or rather to which degree). In the last section I return to the recurrent structural hurdle that LVC needs to overcome in order to build its own model, while engaging in an international structure where another political culture prevails.

e Working against the dominant culture

Chapter 2 showed how LVC's political culture is predominantly built on a low degree of institutionalisation (e.g. strong principles of participation, inclusion and consultation). This chapter has further demonstrated how internal accountability is central to building this model. This somehow contrasts with the contemporary discourse of 'good governance' where accountability is often related to promoting of financial responsibility and 'efficient performance'

(Scholte, 2011: 15; see also Tallberg and Ulhin, 2011; McKeon, 2009a). A member of the food sovereignty alliance and pending member of the LVC explained the different logics:

> The dominant good governance and accountability discourse means that we risk drowning in bureaucracy. This spells the death of social movement and so begins the formation of an NGO or political party. For us, good governance means that communication must be founded in our values.
>
> (Food activist, the food sovereignty alliance, interview, Skype, 10.05.2014)

As the member states here, the dominant good governance and accountability discourse may lead to subtle processes of professionalisation of social movements. Observations of the dynamics in the CFS show that peasant delegates often intervene in CFS sessions to remind the other stakeholders in this UN arena that they, as social movements, work from a different culture and logic than the other stakeholders in the CFS.

> We are different. We are social movements; we are not diplomats or NGOs. We need time to consult with our social constituencies. We cannot work in an individual matter. We are not here on our own. We need to use more time to design our own proposals.
>
> (Latin American LVC member, personal conversation, CFS40, Annual Session, Rome, 18.10.2013)

One LVC member explains that the UN is used to working with NGOs who can work in an individual way and are often accountable to donors, whereas social movements must engage in substantial efforts to consult with their constituencies. While caring for values like collectivism, inclusion, consultation and dialogue, LVC's political culture is somehow inscribed in opposition to the dominant logic in which efficiency is the central criterion (Santos, 2012; Ritzer, 1993).[10]

For instance, when social movement activists work 'necessarily slow' towards an outcome and take time to build consensus they work against the culture of 'teleological efficiency' that is inherent in what Weber (1994) describes as the logic of rationality and bureaucratisation.[11] The prevailing working culture, e.g. the processes in the languages, speed and rhythms that prevail in the arena, means a constant pressure on social movement delegates. An LVC member delegated to the CFS expressed this in the following way:

> I am under pressure to be efficient from many sides. I need time to consult with my peers while at the same time I fear not being as efficient as it is expected of me. It all moves very fast in the processes at this level. It is sometimes very difficult, as we need more time.
>
> (European LVC member, CFS meeting, Rome, 14.12.2014)

The dilemmas and tensions presented in this book reveal that LVC is faced with the following question: Should the movement agree to spend time and energy to develop texts that comply with UN standards and deadlines? Or, should it spend time on consulting and building up the strategy internally? According to Paul Nicholson, 'It is a constant balance between the structure of Rome[12] and the structures of La Vía Campesina. Should we decide to be less UN inefficient – which means more efficiency for us?' (Personal interview, 30.10.2014)

In the latter case, this may be at the expense of a certain international 'efficiency' required to participate at the international policy level. Paul Nicholson, suggests how the movement should ideally prioritise:

> If efficiency means that we are moving too fast without time for the debate, it is not democratic. In this sense, I think we need to run the risk of being ineffective as this is most effective to us.
>
> (Paul Nicholson, Skype interview, 30.10.2014)

One of LVC's internal debates is how to maintain political power at local, national and regional levels so as to keep the movement's horizontal structure, albeit at the expense of a certain international efficiency (Nicholson, 2012: 2). This means that those LVC members who have been appointed as delegates to the CFS are constantly faced with the challenge to negotiate this tension and do the tightrope walking in practice.

The next section of this chapter develops the argument that if we wish to better understand how social movements seek to negotiate their engagement in multiple scales and venues, we must look beyond dichotomies.

4 Beyond dichotomies

LVC presents itself as a radical social movement that has a radical alternative to the neo-liberal model (LVC, 2008). Based on the observations above, we return to a central question of this book: Is it possible to remain a radical social movement while moving into the orbit and state bureaucracy of a UN Committee? Following the predictions of most social movement theories behind the 'classical path', the answer would be no. Social movements' internationalisation is supposed to follow a pattern where leaders, while moving into an elite arena, will transform into a 'cosmopolitan elite' driven away from their base (Tilly, 2004: 154; Tarrow, 1998: 134). Against this dominant view strongly connecting internationalisation with a pre-determined negative path of internationalisation, other scholars suggest that social movements will become stronger from internationalisation. For instance, Gaventa expresses that:

> Transformative, fundamental change happens, I suggest, in those rare moments when social movements or social actors are able to work

> effectively across each of the dimensions simultaneously, i.e. when they are able to link the demands for opening previously closed spaces with people's action in their own spaces; to span across local and global action, and to challenge visible, hidden and invisible power simultaneously.
>
> (Gaventa, 2006: 30)

As stated in Chapter 2 of this book, the rapid processes of neoliberal globalisation and the agribusiness model in the 1980s meant that the global level increasingly impacted the lives of farmers. This led to an increased awareness of the need for peasants to get organised and mobilise at all levels. The need for peasants to engage in a 'multi-pressure environment' was clearly expressed by an LVC member during an evaluation session of the CFS:

> The system is too complex to only engage in at one level. We must engage on all fronts simultaneously. We cannot put all our eggs into one basket. But we must already remember that we have more power if we are active everywhere. We must be creative to find our way to gain more power here and never forget what our vision is.
>
> (European LVC member, CSM evaluation meeting, 16.10.2015)

In line with Gaventa's (2006) social change strategy, the hope is that social movements who engage in the CFS (i.e. experiment with engagement at the global level) will become stronger by developing a methodology than enables them to deal with pressure from different sides and engage in several sites simultaneously: 'We have much more power than we show in negotiations or in the streets alone. We must be better at building our vision by using the different experiences and expertise we have as a growing movement' (LVC member, interview Rome, 16.10.2015).

As the LVC member states here, instead of drawing on the differentiation between 'inside' versus 'outside' categories, it may be more beneficial to explore how to better build synergies between different forms of political activism than to build a common project directed to large-scale transformation.

a ***Radical insiders*** **and** ***grassroots activists***

In the following section, I return to the question: Will social movements automatically lose their radical edge if they take the struggle 'inside' an institutional arena? To explore this question empirically requires reflecting on what notions such as 'inside' and 'radical activism' mean. For instance, we may ask whether the formulation of alternative proposals directed to building food sovereignty policies to build a radically different approach to industrial agriculture is necessarily less radical than other forms of political activism. As one activist engaged in the CFS expressed it:

> When we as a social movement accept to participate in the formal discussions with state actors does it mean that we will 'sell out'? I ask these questions

> because I find myself engaged with the system at the highest level whilst being part of a grassroots social movement. The source of engagement for most food activists is the same, namely injustice. We must find ways to turn this state of indignation and injustice into concrete proposals for positive solutions and alternatives. It is pretty hard to be positive, but easy to get angry about climate change, land grabbing, corporate land grabs, ecological devastation, and so on. We must not give up. We must keep the focus on our struggles to build resilient food systems and address the systemic failure that perpetuates them all.
>
> (CSM member, applying for membership of LVC, interview, Rome, 15.10.2014)

Social movements do not automatically renounce their contentious stances and radical struggle while taking part in international negotiations. Such observations challenge Michels' 'iron law of oligarchy' ([1915] 1962) stating that social movements, like members of political parties, in order to survive as organisations or as social movements increasingly pay attention to adapting to the environment rather than to their original goals of social change. An activist engaged in the CFS for the first time expressed how he experienced that activists perform different and sometimes multiple tasks, which are, in fact, part of the same political struggle:

> Being engaged does not mean that I cannot be an angry outsider as well. Why can't we be both? I am working here [at the United Nations] to ensure that we can be both positive and angry – in fact you are often both positive and angry when you work inside the system.
>
> (CSM member, applying for membership of LVC, interview, Rome 10.10.2014)

LVC seeks to build coherent resistance and a radical alternative against the dominant neoliberal model of agriculture, which to a high degree compels us to think beyond the 'inside – outside' categories (Smith and Wiest, 2012: 161). The following section develops this argument.

b Beyond inside–outside categories

Activists engaged in 'direct activism' (Tarrow, 1998) are for many reasons more visible than those engaged in 'insider activism' within global institutions using a great part of their time to lobby, advocate and draft policy documents. This tension was registered in a discussion within the CSM about 'how to better provide feedback to grassroots'. In this debate one LVC activist, rather frustrated about the framing of the debate, intervened with the following statement: 'This debate is crucial, but please stop talking about grassroots as them out there. I am here, but I am a grassroots activist, too!' (Observation, CSM 'wrap-up' meeting, Rome, 14.10.2014).

Fieldwork observations show that performing different forms of activism does not mean that members of the movement cannot be sympathetic to their colleagues who are also 'angry on the inside'. As one food activist put it in an online discussion among food activists following a CFS session: 'It is good to see that we have people to keep the fire on in there' (Online discussion, Facebook, 04.12.2014). In addition, so-called 'inside activism' in global institutions can take various forms. Such actions may range from giving a press conference, doing a tactical symbolic event, or participating in negotiations. As one LVC staff support person explains it based on his experience following LVC in several international negotiations:

> There are many ways of engaging inside different systems. This is not as black and white as many think. The terms we use are not always adequate to describe what is going on and the different forms of actions that we handle in strategising.
>
> (Skype conversation, 01.07.2104)

Fieldwork shows that social movements and their allies from within the autonomous civil society spaces continually discuss their own role in the CFS and how to build these global practices so that they better draw from social movement struggles around the world:

> We must constantly try to find better ways to work here. We must remember that we have much more power than negotiation of text here. How can we better bring with us here different forms of activism that we have learnt in our struggles around the world?
>
> (LVC member, CSM evaluation, Rome, 10.10.2015)

Given the high level of consciousness and reflection registered when movement activists engage in the CFS, observers should be careful when defining dichotomies such as insider vs. outsider, radical vs. reformist, global vs. local (Massicotte, 2010; Smith and Wiest, 2012: 161).

In order to explore the possible reconfigurations arising from the social movements' experimenting with action across multiple venues, it seems to be important for scholars to recognise the complexity in strategic thinking (Jasper, 2004) and that social movements' resistance and alternative building take place in different venues.

Aiming to better understand the experiences and possible synergies arising from different social movements' experiences with building synergies between different forms of activism, I suggest that observers and activists alike should be cautious not to automatically see some forms of activism to be less 'radical' than others – or at least be clearer about the meaning and consequences of using these terms.[13] The challenge seems to be to build a process of pooling of ideas, and exchange of practices that may complement rather than compete with each other.

The following section presents a final example that demonstrates why it may be helpful to apply a more open lens to the analysis of social movement engagement in a global policy arena.

c ***Passionate politics***

The motivation to engage in a social movement usually comes from great dissatisfaction with the current system (Touraine, 1978). Emotions (e.g. anger, solidarity, outrage, hope) often function as a vehicle for people's mobilisation against social, environmental, and economic injustice (Yang, 2000; Jasper and Poletta, 2010). LVC members who have been appointed as delegates to the CFS often express their strong political and emotional commitment to the cause for which they are fighting. It means that peasant delegates often bring an emotional charge to the negotiations:

> It is a struggle to stop land-grabbing and other pervasiveness in the global food system. It is about policies that keep millions in hunger. Building a resilient food system is about changing lives for the better for millions of people around the world. Our fight is not only a fight over words and then we can go back to the office as usual.
>
> (East European LVC member, interview, CFS41, Rome, 16.10.2013)

Such personal experience of how agricultural industries are causing deep conflicts, displacements, and violent evictions of indigenous families sometimes make the issues at stake difficult to address with a cool head. In the CFS arena this is portrayed by some rural activists, who at times find it hard to control their emotions. As one newcomer to the CFS admitted when I asked him how he felt sitting in a plenary session for the first time:

> It is difficult to sit quietly and listen to the governments talking about investment. The farmers of the world need to hold me back as I cannot always control myself (…) All this bureaucratic talk when I'm just here to stop land-grabbing from happening in my country every day. The governments are bureaucrats. They only see the world out of their office windows.
>
> (East European LVC member, interview, CFS 40, 14.10.2014)

Vera Taylor writes in her analysis of movement strategy within feminist organisations that: 'emotions provide the "heat" so to speak that distinguishes social movements from dominant institutions' (1995: 232). However, when the starting point for peasant delegates is strong personal and emotional experiences, this does not mean that they are devoid of rationality or strategic action. As Jasper and Polletta note in their study of social movements' collective identity, it would be too straightforward to label social movement delegates to be either 'soft-hearted' as opposed to 'hard-headed' (2001:15). This resonates with the main argument presented in this book, namely that we need to be careful not to lock social movements too firmly into institutions. These dynamics arising from observing LVC engagement in the CFS show that this engagement does not automatically mean that they are devoid of strategy action: engagement is overall emotional, angry and strategic all at once.

5 Conclusion

This chapter has demonstrated how LVC leaders engage in CFS in a way that does not automatically correspond to the main predictions of the 'classical pattern' of institutionalisation. The expectation is that social movement leaders engaging with established politics will transform into a cosmopolitan elite resulting in the destruction of local accountability (Friedman, 1999: 40) and de-radicalisation (Tilly, 2004: 155; Tarrow, 1998: 134).

Against this model where a global movement of elites of social movements are gradually becoming cut off from their members (Pleyers, 2010: xxx), LVC actively seeks to build another model. This is registered in the attempts to build a model of internationalisation while retaining a low level of institutionalisation (Chapter 2).

While the risk of elite formation and de-radicalisation obviously remains present within a social movement that is engaging with a UN Committee where another efficiency logic prevails, the movement's reflective nature seems to be an important safeguard mechanism against some of the risks related to the 'classical paths' of institutionalisation. Exploring the practices and attitudes of peasants delegated to the CFS reveals a strong adherence to internal democracy, expressed by the openness and willingness to self-assess. The discussions presented earlier about autonomy and rotation, and how to better build linkages to their constituencies while delegated to global policy work are examples of the ongoing reflection about remembering the struggle on the ground.

To what degree LVC can retain its political culture anchored in a low level of institutionalisation while engaging in international structures like the CFS is identified as a key challenge for the movement. Such complex issues cannot be examined without further identification of the actor–structure relationship between the movement and the UN institution (see Chapter 5). The following chapter debates some of the more pertinent aspects that contribute to LVC's model of internationalisation, namely how to forge alliances with others.

Notes

1 Another example presented by Pleyers (2010) is ATTAC-France. This international network, among others, has been criticised for its lack of internal democracy and in 2006 it was accused of internal election fraud (146–149).
2 The response to the criticism was to develop into a more decentralised, open thematic consultation process, decentralising the site via the polycentric Forum of 2006 (Bamako, Caracas and Karachi).

3 In her book, Desmarais (2007) alludes to the fact that LVC's building of democratic leadership has also been a challenge faced by LVC's own members.
4 With 'reflexive' I refer to the capacity (Giddens, 1984) as well as the will (Pleyers, 2010) of social actors to reflect on their own actions.
5 For instance, fieldwork conducted with different members of LVC in Brazil (2013–2014) showed how Internet remains expensive and limited in the countryside in Brazil.
6 For a discussion about delegation in social movements, see Pleyers (2010: 42–48).
7 For the central issue of trust within La Vía Campesina, see Chapter 2.
8 This is a central point in Santos' text 'Public Sphere and Epistemologies of the South' (2012) where the author presents the concept 'intercultural translation', which is understood as a procedure that allows for 'mutual intelligibility' among the diverse experiences from the Global South and the North.
9 This challenge was communicated to me by an LVC member engaged in the CFS (Conversation, Rome, 16.10.2014)
10 In his book *The McDonaldization of Society* (1993), the American sociologist George Ritzer uses the analogue of the world's largest fast-food restaurant to describe efficiency as a general process of rationalisation where efficiency is understood as the optimal method for accomplishing a task; i.e. 'the fastest method to get from point A to point B' (Ritzer, 1993).
11 This logic of rationalisation is inscribed in what Max Weber called the 'iron cage', e.g. the logic of rationalisation and bureaucratisation and the behavioural structure of organisation inherent in social life in a capitalist society (Weber, 1994).
12 Besides the CFS which is hosted at the FAO in Rome, also the UN based food agencies, the International Fund for Agricultural Development (IFAD) and the World Food Programme, are based in Rome.
13 For instance, what does it means for an LVC member spending personal time and resources to engage as a delegate in laborious UN governance work and then face criticism that this is not 'radical activism' or a 'waste of time'? Previous fieldwork conducted within the alter-globalisation movement reveals that scholars and activists alike face a challenge to nuance the debate and be clearer about what is actually meant by these terms as inside vs. outside, radical vs. reformist.

7 Main tensions and debates

Strategic engagement with others

This chapter discusses some of the main debates and tensions related to LVC's building of its pattern of internationalisation while engaging with others. It shows that the movement does not uncritically follow a pattern of aligning with elite allies to secure its survival. Rather, the movement constantly seeks to assert its autonomy while engaging with others (Borras, 2004: 16; Desmarais, 2007: 123). In order to show some of the challenges and opportunities related to this engagement with others, I present a few examples of LVC's relationships with international institutions, state actors/political leaders, NGOs and academics. This chapter reveals that these relationships are not static but negotiated, debated, and evaluated within the movement.

Whereas the previous chapters of this book mainly focused on the dynamics related to LVC's engagement in the CFS, this chapter brings a number of voices from LVC members who are not directly engaged in global policy work to the forefront. This chapter demonstrates that LVC is simultaneously an actor and an arena consisting of exchange and debate among national, sub-national and regional peasant and farmers' groups (Borras, 2004: 3). The chapter concludes that the fact that LVC remains a movement with room for internal debates and tensions is important for its cohesion and evolution. This, I suggest, may particularly be the case when LVC moves into complex global policy terrain.

1 Alliance building and autonomy

LVC faces the constant challenge of needing to build alliances that may help advance the movements' agendas and goals while simultaneously seeking to preserve its autonomy (Borras, 2004: 3; Desmarais, 2007: 123). It means that LVC is in an ongoing process of identification and clarification of what constitutes an alliance and who the most suited allies may be (Desmarais and Nicholson, 2013: 6). The overarching framework of LVC on the issues of alliances and autonomy is clarified in some of LVC's policy statements:

> We do not have a choice as to whether we interact with others who are engaged in our arena – but we have a choice on how we work to effect

> the changes we desire (…) Our efforts to defend peasant agriculture/culture and rural areas cannot succeed without cooperation with others. Where we share objectives and can join forces over particular issues with another organization the Vía Campesina should enter into strategic alliances. Such alliances must be politically useful, carefully articulated in a formal agreement with a specified timeline and mutually agreeable. The Vía Campesina must have autonomy to determine the space it will occupy with the objective of securing a large enough space to effectively influence the event.
>
> (LVC, 2000, cited in Borras 2004: 16)[1]

Paul Nicholson argues that LVC's struggle for autonomy has historically been fought on two main fronts: a) With (inter)governmental institutions and b) with NGOs. In order to define autonomy, I refer to the definition provided by Borras (2008), scholar-activist and ally of LVC, who distinguishes between two concepts: 'independence' and 'autonomy':

> Independence is often seen as a choice in 'absolute terms' – groups either allow themselves to be co-opted by these international institutions, or they do not, and are thus insulated from any form of external interference or influence. By contrast, autonomy is 'inherently a matter of degree' and refers to the amount of external influence in the agrarian movements' internal decision-making. In this view, an organisation may have relationships with other entities, but what matters is the terms of those relationships.
>
> (Borras, 2004, based on Fox, 1993: 28)

Members of LVC often emphasise their concern over retaining a high level of autonomy, whether it means responding to an invitation to a meeting or a partnership agreement, holding a political meeting, making decisions regarding a funding strategy or situations where the movement engages with other civil society organisations. This implies asking questions such as: Where should we engage? On whose terms? With whom? How? To do what? (Paul Nicholson, Skype interview, 01.07.2014).

The following sections present some of the challenges and opportunities for LVC in its attempt to combine a high degree of autonomy while engaging with carefully selected international institutions.

a La Vía Campesina and international institutions

Whereas LVC is open towards engagement with some international institutions, it categorically rejects any invitation to be a partner, enter into dialogue, or engage in consultative processes with international financial institutions, such as the World Bank, IMF, WTO (LVC, 2004a: 117; Borras and Franco, 2009: 13; Desmarais, 2007: 29). LVC clearly takes a confrontational stance towards these institutions, as they are the epitome of what the movement opposes:

undemocratic institutions that are often seen as tools of neo-liberalism (LVC, 2004a: 118, Borras and Franco, 2009: 13; Desmarais, 2007).

As Chapter 3 showed, the UN has historically been seen by LVC as a more democratic institution than other international institutions due to its 'one country – one vote system' (LVC, 2012a). This explains why LVC is more open to building relationships around common areas of work with different UN bodies, such as the Food and Agriculture Organization of the UN (FAO), and the International Fund for Agricultural Development (IFAD) (Borras and Franco, 2009: 11).

For instance, LVC has repeatedly stated that one of the main motivations behind their increased engagement with the FAO has been to struggle for positive change in an institution that could become a counter reference to WTO. As early as 2004, given the political-economic context in which agriculture became increasingly dominated by the emergence of the agribusiness industry, LVC advocated for more power to be transferred to the FAO in terms of decision-making on agricultural policies (LVC, 2004a: 3). This was, however, just one month before the highly contested FAO report on biotechnology came out (Müller, 2011). LVC criticised this report for supporting the biotechnology industry and for embracing GMO agriculture 'as a threat to people's food sovereignty' (LVC, 2004b). During its 4th International Conference, held in Brazil in 2004, this episode propelled the movement to launch an official press release that stated that the FAO had 'joined the IMF, WTO and World Bank as the guardians of capital' (LVC, 2004b).

The delicate relationship between LVC and the FAO is also portrayed by the heated debate on 'land grabbing'.[2] In October 2012, Jose Graziano da Silva, Director General of the FAO called for a 'Sheriff' to regulate land acquisitions (*The Guardian*, 2012). However, this was the very same year he co-signed an article with the European Bank for Reconstruction and Development (EBRD) that, according to LVC, was a way of 'promoting the destruction of peasant and family farming' and calling on governments to 'embrace the private sector as the main engine for global food production' (LVC, 2012d). LVC interpreted this as a step closer towards the business sector, and aroused anger among members of the organization. This was rather unfortunate for the FAO, at that time in a process of clarifying the terms of a relationship through an 'exchange of letters'[3] between the peasant movement and the FAO.[4] These examples demonstrate that LVC's relationship with selected UN institutions is far from static. It shows how relationships can quickly disintegrate if members of the movement feel that trust has been broken.

b Institutions as arenas

Various scholars have reiterated the need for social movements to build political support for their struggles by establishing relationships with influential allies inside institutional arenas. Relationship-building with 'institutional brokers' potentially enables actors to urge political authorities to modify their policies by taking into account the claims of societal movements (Giugni and Passy, 1998: 6; Kitschelt, 1986; Kriesi et al., 1995; McAdam, 1982; Tarrow, 1994; Tilly, 1978). A great deal of social movements' substantial impact and influence depends on the 'mediation' by political allies willing to take up the movements' claims in institutional arenas and, therefore, social movements also rely on the presence and availability of political alliances (Kriesi et al., 1995; Tarrow, 1994; Della Porta et al., 2006).

Building a relationship with allies inside institutions is a central part of LVC's lobby strategy, in particular when it engages with the UN's specialised agencies. Data from fieldwork conducted in the FAO house[5] show that representatives of peoples' organisations – in parallel to engagement in CFS work – often meet with FAO employees (on both a formal and informal basis) in their search to influence the political and strategic agenda of the institution.[6] As the examples presented in this chapter illustrate, UN institutions are – like states and social movements – not only actors but arenas or sites of tensions and internal debates. As Borras explains: 'UN agencies are comprised of various actors that have different and, at times, conflicting and competing agendas, some of which may support LVC's agendas at different times, others not' (2004: 19). Some UN officials have strong relations to the private sector, while others have a history with social movements, which may make them more inclined to embrace social movements' proposals (McKeon, 2009a; Borras, 2004). Consequently social movements may strategically use UN arenas to identify and exploit contradictions within (international) institutions. During its 2008 International Conference in Maputo, LVC became more open towards the idea of building institutional alliances than it had been in the past:

> New institutional alliances may create space for new policies. It is up to La Vía Campesina, and its alliances to closely analyse the situation and be prepared to step up to the plate when there is something to gain for La Vía Campesina.
>
> (LVC, 2008: 13)

In parallel to LVC's interaction with the UN's leading food security institutions based in Rome, LVC is engaging with other bodies in the UN system, such as the United Nations Development Programme (UNDP), The United Nations Conference on Trade and Development (UNCTAD) and the UN Human Rights Council (UNHRC). In the case of the UNHRC, LVC played a prominent role in developing the proposed United Nations Declaration on the Rights of Peasants and Other People Working in Rural Areas. It played a key role in drafting and negotiating the terms for the established open-ended intergovernmental working group (Claeys, 2013; Edelman, 2012). A rather different pattern can be traced when we assess LVC's engagement with other UN intergovernmental bodies and processes.

Research conducted during the 'Rio+20 People's Summit' held in Rio de Janeiro in June 2012 and later the same year at the UN Headquarters, New York, revealed the critical stance of many civil society activists in the process of developing the Sustainable Development Goals (SDGs). LVC leaders familiar with UN processes interviewed for this book reveal how members of the movement criticized the nature of the consultative participation in place to engage CSOs in the inter-governmental process. For instance, LVC representatives participating both in the parallel 'People's Summit' and the official UN Rio+20 negotiations in particular emphasised the problem of bringing all categories of farmers together in UN processes, expecting small-scale farmers to find common ground and to speak with only one voice (LVC activist, Rio Peoples' Summit, 15.12.2014). As discussed in Chapter 2, in the eyes of many LVC activists, the tendency to lump actors together in UN processes is a sign of how differences are ignored. A frequent criticism in this regard has been the lack of clarification on 'who speaks for and who represents whom' as well as the timeframes set by the UN without substantive time for grassroots consultation. This explains why LVC has sometimes chosen to walk away from these types of processes (US based LVC member, interview, 16.05.2013).[7]

The ambivalence toward prevailing UN processes was reflected in LVC's internal debate over whether the movement should participate in the official UN Rio+20 process or choose to walk away. Nettie Wiebe explained to me how LVC strategically decided to appoint a few members to engage in the official UN process from the rationale that it would be better to be present rather than leave the industrial agribusiness to be the only voice of farmers at the UN conference (Nettie Wiebe, Conversation, Jakarta, Indonesia, 12.06.2013).

The different standpoints that LVC takes towards different UN processes confirms that the UN is a rather *diverse* terrain for collective action, as it is constituted by different UN forums, processes, agencies, staff personal, etc., with different mandates, organising principles and normative values (see e.g. McKeon, 2009a; McKeon and Kalafatic, 2009). LVC's insistence on defining the terms of the processes in which the movement may choose to engage shows that the movement is not static. Rather, it strategically decides when to engage and disengage (Desmarais, 2007: 121). As Desmarais has observed, the active choice of not engaging may in fact be an effective strategy for the peasant movement to delegitimise the institution (2007: 37).

The following section presents how the social movements' relationship with the state/political leaders remains one of the main debates within the movement.

c Ambivalences towards the state

The strategy of combining mobilisation with some aspects of institutional engagement and building relationships with states and other institutional actors is a heated debate among social movements and scholars alike (Kriesi, 1996, Kitschelt, 1986; Tarrow, 1994; Piven and Cloward, 1977; Schumaker, 1975; Gamson, 1990; Della Porta and Diani, 1999; Borras, 2004; Borras and Franco, 2009).

For instance, Pleyers shows that social movement activists often express a great degree of ambiguity towards the state and institutions (Pleyers, 2010: 222–224). In his investigations of the alter-globalisation movement (2010–2010), in which LVC has played a central role, Pleyers illustrates how activists tend to strongly oppose the state and/or institutions in their discourses, but in practice often look for their potential protection and support (e.g. subsidies) (2010: 223). Borras contends that LVC was born in the context of

> the state's partial withdrawal from its traditional obligations to the rural poor and the waves of privatisation that affect poor people's control over natural resources and access to basic utilities have also left many poor peasants and small farmers exposed to the harshness of market forces dominated by the global corporate giants.
>
> (Borras, 2004: 1)

The socio-economic context of the neo-liberal ideology, leaving minimalistic room for state interventions, explains why the global peasants' movement has often reiterated the need to 'widen policy space for the nation state' (Borras, 2004: 1). In a similar vein of thought, Van der Ploeg discusses the critical role of the state as a 'regulator' that must win authority over capitalist market regimes so that the welfare state and people's agency can flourish (2008: 239). Yet, in practice, the state often remains a double-edged sword for social movements (Pleyers, 2015; McKeon, 2015). On the one hand, states promote narrow and short-sighted objectives; on the other hand, states can function as the building block for the protection of citizens' rights within a democratic framework (Mann, 2014: 54).

Pleyers argues that states and institutions are often simultaneously opponents and allies in the movement struggle, yet that ambivalence is often left out from the discourse of social movements (2010: 223). From this viewpoint, the state cannot be identified merely as an enemy or ally of a social movement: 'State agencies may be either allies or opponents: Government agencies can support or oppose movement claims, since some of the agencies might believe in movement goals and others hold opposing beliefs' (Gale, 1986: 205).

Like social movements and institutions, states are arenas themselves and comprise different ministries and departments, with at times different and conflicting aims and agendas (McKeon, 2015: 65; Weiss, 2012; Randeria, 2007). As demonstrated in Chapter 4, which explores some of the dynamics arising from social movements engaged in the Committee of Food Security (CFS), peasant delegations and other activists actively exploit such contradictions when they seek to influence government representatives during and between CFS sessions: 'Governments can choose to be on our side. If they dare to say no to industrial agriculture and collaborate with us we can strengthen peasant agriculture and produce even more and better food' (LVC member, personal conversation, CFS-rai negotiations, Rome, 09.08.2014).

While lobbying and advocacy work are central parts of the strategy to win support for their policies, social movement activists observed in the CFS sometimes express their feeling of mistrust while engaging with government representatives. Such ambiguities are often anchored in different experiences on the national policy scene. This explains the mix of hope, optimism and disappointment with popular/progressive governments and political parties (Wittman, 2012: 10; see also Rosset and Martinez, 2010).[8]

The official position of LVC is that the movement has to remain independent from political parties, but LVC member organisations have the autonomy to decide whether and how to engage with the national policy arena (Vieira, 2011; Rosset and Martinez, 2014: 150). It is up to each member organisation of LVC to decide who they engage with and how in their national and regional contexts, and LVC as a global movement is normally open to engaging in dialogue with progressive governments that are sympathetic to the movements' agenda. In the next section I will delve into this collective strategy to maximize pressure and influence and show that it stems from the logic that negotiations must ideally be combined with mobilisation (Borras, 2008: 114).

d One foot in, one foot out

Borras describes how LVC's collective strategy ranges from mobilisation to negotiation. This strategy consists of a dual inside–outside pressure of a) mobilising and demonstrating in opposition to the policies and institutions that are hostile to the movements' interests in order to prevent or change them, and b) negotiating and collaborating in order to influence policy changes (2008: 114). Borras emphasises the dialectic potential between these forms of political activism: 'While in official discourse, these strategies often appear to be separate, competing or even conflicting, ideally they should be linked' (2004: 22).

The strategy presented above is also known as the 'twin track' strategy (Borras, 2004; Fox, 2000) or a 'one foot in, out foot out' strategy pursued, among others, by several members of LVC in Brazil. One of the members and supporters of the Landless Rural Workers Movement (MST), Álvaro Delatorre, explained the significance of this strategy:

> In the MST we have the saying 'pau e pauta' meaning that you need to have dialogue and to negotiate. Social movements are not the ones who distribute land to the rural population or construct better infrastructure for children to go to school. We need to negotiate for better policies to improve life of the marginalised in our societies. This does not mean that we do not remain a radical social movement. Negotiation is a part of the very raison d'être of a social movement.
>
> (Álvaro Delatorre, member of MST, coordination in the State of Rio Grande do Sul, MST member, Rio Grande de Sul, interview 30.3.2014)

Since 1984, MST has challenged the lack of land reform in Brazil by occupying vacant arable land owned by wealthy landlords and redistributing small parcels to landless workers. MST claims to have 50,000 families living in long-term, legally recognised settlements, while another 90,000 members are living more precariously in camps on contested property and have gained access to more than 15 million acres of land.[9] While MST is mostly known in Brazil for its militant occupation of factories, ministers and members of MST interviewed for this work emphasise that mobilising cannot be separated from the politics of negotiation, expressed in the movements' effort to combine designing and influencing public policies in concrete ways with powerful mobilisations (Wolford, 2004; Wittman, 2009). For MST, these are two sides of the same struggle:

> We must work with the political leadership to transform agendas into political priorities. There is a strong sense of responsibility of the members: We must negotiate to ensure social infrastructure for the people staying in the country with education, water, electricity, irrigation systems, appropriate technology to small agricultural needs, and win support to our cooperatives. However, we should never think that social change follows from negotiations only. Our strength is the ability to mobilise. Without this mobilisation there will be no dialogue with governments.
>
> (Álvaro Delatorre, MST, Santa Maria, Brazil, Interview, 15.03.2015)

Thus, instead of talking about either/or strategies, the movement seeks to build synergies between these different forms of political activism, which include political dialogues. In an interview, Delatorre explained that this, however, is not without tensions. The MST veteran shared details with me on how, whilst the movement was holding its National Congress in Brasília, President Dilma went to the State of Mato Grosso in Brazil to celebrate the agribusiness industry:

> We clearly denounced Dilma's speech, which openly applauded the agribusiness industry. While Dilma was sat on a tractor in Mato Grosso posing for the media, we mobilised approximately 15,000 people in the streets of Brasilia to protest against agribusiness. We showed the power we have in the streets. The next day Dilma called the MST to talk to us.
>
> (Álvaro Delatorre, MST, Santa Maria, Brazil, Interview, 15.03.2015)

While some members find it important to influence policy-makers and increase dialogues, the degree to which the MST should interact with the political elite remains an ongoing internal debate. The ambiguity towards political leaders was expressed during the 2014 National Conference: ahead of the movement's 6th National Congress in 2014, MST activists debated whether they should invite President Dilma to their National Congress in February 2014. Despite not inviting the President of Brazil to the National Conference, during the

Congress, members of the MST declared their support of the president in 2014 presidential elections.[10]

Field observations conducted with a variety of actors in different regions of Brazil reveal the rather permeable institutional infrastructures, where farmer's organisations have carved out different ways to engage directly in national policy processes and programmes. In an interview conducted with the Brazilian Minister of Agrarian Development, Pepe Vargas, the Minister explained how the long tradition of direct dialogue with peasant organisations is expected to lead to a more sustainable model of agriculture.

> I would say that mobilised peasants, and in particular members of LVC from the South to the North, help to shape the values of our development model. Participation of the organisations made by and for citizens is central to develop the policy programmes that should serve them. Permanent dialogue is not only crucial for sound public policies. A society where politicians do not listen to all voices of society is not a democracy.
>
> (Pepe Vargas, Brazilian Minister of Agrarian Development, 2012–2014, interview, Brasilia, 25.03.2014)

The debate on participation and inclusion/exclusion in policy dialogues in Latin America must be seen in the context of a history of dictatorship and political authoritarianism where social movements did not have access to dialogue or negotiation tables, as decisions were made by the political elite and political parties behind closed doors (Gwynne and Kay, 2004: 202). The transition from authoritarian regimes to democracy in many Latin American countries opened new possibilities for social movements. In this regard, Brazil has often been seen as a pioneer in providing policy space for direct civil society participation in policy dialogues and programmes (Dagnino, 2002; Fung and Wright, 2001).[11]

This tendency of activists from social movements gaining access to decision-makers in different types of 'new hybrid' democratic spaces (Cornwall and Coelho, 2007) has triggered critical reactions and analyses from scholars and activists themselves (Dagnino, 2010; Dagnino and Tatagiba, 2010; Schuurman, 1993). For instance, Schuurman (1993) points out that the new access of Latin American social movements to decision-makers in the transition to democracy led to a tendency to incorporate social movement leaders, but not their demands, with negative consequences for popular groups. The awareness of such risks may explain why some activists decide not to prioritize to engage in institutional politics, lobby and advocacy work but instead strive to have an impact on the transition by shaping the political culture (Wampler and Avritzer, 2004). This example illustrates well the ongoing debate on opportunities and risks related to engagement with governments and institutions.

Moving to the other side of the globe to the 6th International Conference in Jakarta, interviews with LVC members demonstrated that others activists have indeed embraced the 'dual strategy' referred to in the section above. For instance, youth members of the national farmers union in Jakarta explained how different kinds of political activism must complement and reinforce each other. In a conversation, one young member emphasised that lobbying and advocacy work are important tools to gain influence and leverage, but only as long as these are followed by 'direct activism'. In order to illustrate his point, the activist shared with me pictures of him and other SPI members blocking the way for bulldozers in their local villages, resulting in violent confrontations with the authorities. The crowd of young SPI members, gathering around me during this conversation, eagerly showed me their photos of Henry Saragih, former Secretary General of LVC, climbing on a bulldozer in a conflict over the palm oil industry. Before heading into the next session of the Peasant Assembly, the members emphasised the importance of this type of activism as the most efficient social change strategy to exert 'pressure from everywhere':

> Merely relying on a strategy to influence policy in the public arena is simply not enough to push for fundamental social and political change. In order to have a significant impact, we must continue to carry out our coordinated mobilisation and find ways to do political advocacy work. We must strive to influence policy in every space where decisions are made. The political elite does not always listen to us but we continue to put pressure on them until they start listening. Today, we have a worldwide movement behind us and governments start to realise that they cannot ignore us anymore.
>
> (Youth member, SPI, LVC's 6th International Conference, Jakarta, 15.06.2013)

3 Towards the implementation of political visions

a Building national food policies based on food sovereignty

Today, more researchers express interest in exploring how rural organisations can use international cooperation and building of alliances with decision-makers. Social movements and academics alike strategically use these types of relationships to push to build support for their visions (e.g., Claeys, 2013; McKeon, 2015; Whiddon, forth.). In this regard, a particularly noticeable trend is how an increasing number of countries are in the process of adopting a framework of legislation for agriculture, food, and nutrition that enshrines rights-based principles of entitlements and access to food in national policies, public policies, laws and constitutions (Beuchelt and Virchow, 2012; Claeys, 2013). Constitutional recognition of the right to food sovereignty has been achieved in a number of countries, including Ecuador, Bolivia, Nepal, and Venezuela, and other countries are following suit.[12] Legal scholars have noticed how the mobilisation of civil society actors and social movements has played a key role in supporting the last few years' of legal developments (De Schutter, 2013b: §53; Claeys, 2013). While access to policy dialogues differ from region to region, today, peasants' organisations around the world, along with indigenous movements in Latin

America, make an increasing contribution to shaping major public debates and changing government policies, laws and new constitutions (Cortez, 2009; Le Bot, 2009). One Latin American LVC leader, engaged in global, regional and national policy while also being an agroecology farmer herself, explained that during the last few years she has experienced a 'changing attitude' among LVC members and organisations.

> We all have our different strategies built on our different experiences. However, I think there is an increasing recognition of the need to combine mobilisation with some degree of political dialogue. The openness towards institutional engagement, I think, is a result of social movements having started to harvest the fruits of building a dual strategy over the last decade.
>
> (Latin American LVC member, interview, CFS, 20.05.2014)

LVC carefully monitors where governments may start to become more open to its proposals. This is portrayed in a policy document from LVC's 4th International Conference in Maputo 2008, stating the following:

> The shifting balance of power may create opportunities in the near future to establish space where we can defend and work on the implementation of our proposals based on food sovereignty. Food sovereignty is now part of the mainstream debate on food and agriculture and its level of recognition is steadily increasing.
>
> (LVC, 2008: 13)

Basque farmer and one of the strategic leaders within LVC, Paul Nicholson, has presented a hopeful view on the 'co-relation of building food sovereignty', that has become more visible after the food crisis of 2008/2009.

> This is the combination of the energy crisis, the food crisis, the crisis of climate change, the financial crisis. This isn't any kind of partial crisis, but rather a crisis of the entire economic and social model. We have to propose a social transformation. It is within this framework that food sovereignty has relevance. But the processes in each continent, in each country, are very different. The objective is not to homogenise these processes. It is clear that the velocity and temporality in each region are very different. But there are regions and countries in the world where institutions are taking on the necessity of food sovereignty and are asking for peasant support for how to apply policies of food sovereignty. This is very new. The co-relation of forces is changing and our discourse has increasing capacity every day. There are some countries where this co-relation of forces is making it possible, like in Bolivia, with Evo Morales, or Nepal, or Ecuador. In Ecuador, La Vía Campesina has participated in the elaboration of the new constitution.
>
> (Paul Nicholson, cited in an interview with Hannah Wittman, cited in Patel, 2009: 680)

Thus, while such institutional processes may be very different from country to country, the aspiration is that collaborations with governments can lead to the development of participatory mechanisms that may help to integrate the right to food in national legal frameworks. This resonates with a central rationale for grassroots movements to use time and energy to engage in institutional processes, namely to gain recognition that can help to open up policy spaces that did not exist before. Paul Nicholson has expressed this sentiment:

> One of the most important things that we have learned while building our movement has been our ability to rebuild our pride of being peasants. Now we are proud to be recognised by major institutions such as the FAO and the Human Rights Council.
>
> (Paul Nicholson, cited in *The Guardian*, 2011)

As Chapter 5 discussed, the increased recognition of peasants in authoritative UN global policy arenas has opened new spaces for LVC to win support for struggles on the ground. The increased openness from international institutions towards LVC does not mean, however, that LVC uncritically engages in new spaces and partnerships. The movement builds on its historical experience (Desmarais, 2007: 11 and 119–121) and remains highly alert to the fact that engagement with the external world brings with it both opportunities and constant risks. This correlation will be the focus of the following section.

b Balancing opportunities and risks

The increased capacity of peoples' organisations to manoeuvre within global institutions and take part in complex deliberations and negotiations, combined with their direct contact with those on the ground, have made them attractive allies for governments, NGOs and intergovernmental institutions. LVC leaders interviewed for this work often express how they experience becoming attractive partners to enhance the legitimacy and support of international institutions and governments. For instance, one African LVC leader explained to me how he and other representatives to the CFS, as soon as they arrive at the airport in Rome receive phone calls from both government representatives and employees of the other UN food institutions in Rome.[13] The member explained that this occurred to the point that he even felt that it was rather 'disturbing' as he had to stay focused on the already demanding CFS processes (African LVC member, personal conversation, CFS 40 annual session, Rome, 15.10.2014).

Thus, while civil society actors use global meetings to build alliances and pressure governments to win support for their political projects, the direction also goes the other way. Consultation with civil society is largely being seen as a benefit that correlates positively with ownership and improvement of the quality of governance (O'Brian et al., 2000; Heller, 2013; McKeon, 2009a). Governments and inter-governmental institutions often directly depend on civil society for their effectiveness,[14] legitimacy and their very existence (Desmarais, 2007: 118). LVC member, Paul Nicholson, explains this tendency:

> The UN needs social movements. This is a part of the revitalisation of the UN. It needs social movements as the real protagonist in order to legitimise its work. We must always be alert. Institutions and NGOs offer appealing partnerships and funding possibilities. We see it more and more. The temptation or risk is that social movements will go for short-term partnerships. We must always seek to analyse the long-term consequences of our engagement. We must ask ourselves such questions as: What kind of engagement? What kind participation or partnership? Which room for conflict? These relations highly depend on carefully analysing what may be gained and what is at risk.
>
> (Paul Nicholson, Skype interview, 30.10.2014)

As the quotation above shows, LVC leaders remain aware of the need to carefully distinguish between a space for 'meaningful participation and a situation where participation becomes mixed up with manipulation', what a LVC member called the risk of 'partipulation' (Conversation, LVC member, Rome, April, 2013).

In particular, the historical experience of others seeking to legitimise their policies and programmes, by claiming that the peasant movement participates, has made LVC alert to the risk of giving legitimacy to processes that may end up going against its interests and/or seeking to dilute or silence the opposition (Desmarais, 2007: 11 and 119–121). In an age where partnership-building and multi-stakeholderism become more and more celebrated, LVC is increasingly faced with the challenge to carefully monitor this rapidly changing trend and identify how to balance the risks against opportunities related to the engagement in the different types of spaces that are opening. Against this backdrop, the following section sheds light on the ongoing tensions between LVC and NGOs (Borras, 2004: 16; Desmarais, 2007: 88–103).

4 NGO and academics: tensions and collaborations

a LVC and NGOs: an ambivalent relationship

For LVC, NGOs remain among some of the most controversial allies (Borras, 2008: 29; Desmarais, 2007: 23–26). The 'love–hate' relationship between peasant movements and NGOs (Borras et al., 2008: 197–198) has been registered on several occasions during the fieldwork conducted for this book. One the one hand, this book (Chapters 5 and 6) demonstrated how, given the recent engagement of social movements in the laborious UN policy processes, they require substantial technical, human and financial resources that grassroots movements like LVC do not always have. This is one reason why alliance-building with more resourceful actors like NGOs remains a key aspect of social movements' engagement in the CFS arena[15]. On the other hand, this book shows the ongoing power struggles within global civil society and LVC's efforts to carve out a space and fill it with peasant voices in a world where dominant NGOs have tended to speak on behalf of 'the poor' and of peasants (Chapter 3; see also Desmarais, 2007: 29). This uneasiness of engagement with NGOs was, for instance, expressed by a social movement activist in a message to other members of the Civil Society Mechanism ahead of the annual session of CFS:

> You expect us to collaborate around building positions with NGOs, but you cannot ignore that NGOs here are perceived as the enemy in our countries. We have a difficult time trusting them around the table here.
>
> (Latin American CSM-spokesperson for the fisher folks, CSM meeting, Rome, 16.10.2013)

This tension between social movements and NGOs must also be understood in the historical context of LVC against the emergence of the so-called 'third sector'[16] and the international elite of NGOs claiming to 'speak for the people' without creating either deep grassroots connections or means for ordinary people to speak through them (Kaldor, 2003: 85; Tilly, 2004: 152). This tension is particularly pertinent in the context of social movements' engagement in the UN arena, where international NGOs have more UN experience, technical expertise and resources available to follow UN processes. This means that there is a constant risk of NGOs ending up 'monopolising the microphones' and taking over the political leadership (McKeon, 2015: 185).

While NGOs do not form a homogenous group, observers have particularly criticised international NGOs for acting on the global policy arena without clear representative mechanisms and without properly consulting with those they most often claim to represent (McKeon, 2009: 177; Tallberg and Uhlin, 2011; Tarrow, 2014). Some activists even criticise NGOs for operating with the same efficiency logic and lack of transparency as IMF and the World Bank (Mann, 2014: 6; Patel, 2009). NGOs are often portrayed as entities that follow a more individualist and hierarchical style than social movements (McKeon, 2009a), and that are less concerned with internal accountability to members (Tallberg and Uhlin, 2011).[17] This is confirmed by a many members of MST and LVC, referring to the preparations of the People's Summit in Rio+20:

> We have seen it so many times. NGOs often act in a very individual manner. Sometimes, they even think they can deliver individual statements and draw up campaigns by themselves and then they expect us to put our name on it afterwards. What kind of collaboration is that?
>
> (Latin American MST/LVC member, interview, Brasilia, 28.03.2013)

Part of the social movements' criticism derives from the fact that NGOs use global meetings to gain global visibility for their own agendas. This was registered, for instance, during one annual session of the CFS in Rome. Here, I observed how one European NGO representative entered the CFS in the midst of a session, got seated beside me, opened his email inbox and took a picture of himself in front of the CSM flag. When he presented himself to me, he explained that this was just 'a short visit to the CFS session' as he had another important meeting in another European capital (NGO participants, CFS41 annual session, Personal Conversation, 13.10.2014). Whereas there may be various ways to explain the conflicts between NGO and social movements, one tension is arguably that different groups are anchored in different interpretations of reality and social change strategies (Pleyers, 2010, Ch. 2; Bringel, 2015: 131).

More resourceful professional NGO employees have a tendency to dominate global UN meetings (McKeon, 2009a: 177) and are often (like most governments) trained to have a strong focus on outcome. While NGOs are often claiming to be more 'efficient' in working towards results, social movements often act from a different logic of transnational activism: one that holds strong values of participation, inclusion, consultation and a process of 'educational, learning, knowledge and mutual enrichment through exchange of experiences' (Bringel, 2015: 131). Without attempting to make a clear-cut schematic wedge between social movements and NGOs and with the dominant working culture of many international advocacy NGOs in mind, NGOs can be placed at a higher degree of institutionalisation in the continuum presented in Table 2.1.

b Towards a more mature relationship?

For LVC alliance-building with NGOs remains both an opportunity and a challenge. On the one hand, LVC can benefit from NGOs' support and analytical skills to address technical aspects of the global system. On other hand, LVC is aware of the constant danger that, to use the words of Pleyers, they will end up becoming 'overwhelmed by NGOs' (Pleyers, 2010: 127). This tension is particularly noticeable in the context of social movements engaging in a UN setting. NGOs have intrinsic advantages in working at this level: they often possess the technical skills, speak better English and are more experienced (and comfortable) with the logic of lobbying and advocacy in the world of UN diplomacy. Such imbalances mean that there is a constant risk that NGOs will end up assuming the political leadership of the CSM group. In this regard, the emphasis placed on the political leadership of peoples' organisations – who encourage NGOs to take on a facilitator role in the design of the Civil Society Mechanism – is remarkable.[18] The priority given to the participation of peoples' organisations, as those categories of society often most directly affected by food insecurity is clearly stated in the CSM founding document.

> If there is a risk that the voices of peoples' organisations and social movements are at risk of becoming crowded out by NGOs, particularly from the North, it may become necessary to apply the CSM quota system in

> order to ensure balanced and equitable participation in CSM working groups.
>
> (CSM, 2010b)

It means that those who dominate the global policy processes, panels and debate in the CSM are faced with a quota system where only one slot of 11 civil society constituencies are reserved for NGOs. Promoting peoples' organisations to be in the political leadership while NGOs are required to take on a supportive and technical role has not gone uncriticised by some NGOs.[19] Within the diverse group of civil society actors engaged in the CSM, it is not only important to build a political awareness of the different roles and mandates. Equally, a large amount of trust and strong bonds of solidarity must be built among members seeking to build the collective space for engagement. An LVC member expressed this in a CSM preparation meeting ahead of the CFS40 annual session: 'We cannot be here all the time to monitor everything; we cannot be in Rome all year. We must trust you are here in Rome when we cannot be here' (East European LVC member, CSM meeting during the CFS40, 14.10.2013).

The determination to build 'fruitful relations between movements and NGOs is not only registered in the CSM space. It was also communicated to me in interviews and conversations during LVC's 6th International Conference in Jakarta (10.06.2013).

During the conference, an NGO member shared with me his reflections from engaging in the movement's group of 'friends and allies' that had steadily expanded during the last years. While only a few NGO representatives were invited to participate at the LVC first International Conference in the parallel forum for NGOs in 1996 in Mexico, around 30–40 allies participated in the 6th Conference in Jakarta. The NGO representative explained this change as following: 'I think that more NGOs have seen the importance of clearly stating our different roles and how it is a strength for all of us to work together' (NGO representative from a European country, personal conversation, LVC's 6th International Conference in Jakarta, 14.06.13).

Building clear relationships between social movements and NGOs can be constructive as it may assist a mutual learning process and the deeper reflections of how synergies or 'cross-fertilisation' between different agendas, horizons and aims can be built (Pleyers, 2010). It also sparks a debate on the different identities, mandates, accountability, and types of organisational legitimacy within the world of civil society (McKeon, 2015: 141–145; Desmarais, 2007: 21–26).

The increased focus on the unsustainability of the funding models[20] pursued by big donor-driven NGOs has led to a legitimacy crisis for some of the NGOs that have played a dominant role in the international arena (McCloskey, 2012; Banks et al., 2015). This tendency was expressed by the global civil society alliance, CIVICUS, in an 'Open letter' in 2014.

> Sadly, those of us who work in civil society organisations nationally and globally have come to be identified as part of the problem. We are the

> poor cousins of the global jet set. We exist to challenge the status quo, but we trade in incremental change. Our actions are clearly not sufficient to address the mounting anger and demand for systemic political and economic transformation that we see in cities and communities around the world every day.
>
> (CIVCUS, 2014)

Such signs of self-criticism by some NGOs show that these relations are not static but rather ongoing processes that are evolving over time (Smith, 2010). As Borras underscores, 'it is not the NGO per se that is problematic. Rather, it is the term of the relationship that matters' (2008: 29).[21]

While NGO representatives who are trained to do advocacy work and lobbying and used to flying from one capital to another[22] are sometimes less reflective of their own roles and issues of internal accountability (Talin et al., 2010) the picture is, however, more nuanced. Neither social movements nor NGOs are homogenous groups that follow the same logic or style of action.

Field observations during meetings and sessions of the Civil Society Mechanism reveal this rather diverse picture. For instance, NGO representatives in CSM working groups[23] who are involved on an all-year basis (not only for yearly events) appear to engage diligently to support the vision of the CSM. In this context, it is worth remembering that the CSM space has largely been built on the practices and learnings gained from the participation of social movements through the International Planning Committee for Food Sovereignty (IPC). As Chapter 3 showed, since 1996 the IPC has been a central space for rural social movements to meet face-to-face, share their practices of engagement, and an opportunity to promote mutual understanding (McKeon, 2015: 29).[24]

While challenges remain for NGOs and social movements to share the CSM space, the CSM is an example of rural movements finding new ways to strengthen the food sovereignty model by building alliances that link peasants' struggles to other sectors such as labour, environmental and indigenous rights movements. Building a model based on complementariness between the political legitimacy and mobilisation capacity of peoples' organisations with the analytic and advocacy skills of NGOs in a mutually reinforcing relationship seems to be a collective task for all civil society activists seeking to put pressure on governments from within international institutions more effectively. This collective process of learning to work together can also be seen as a concrete expression of 'Uniting in Diversity' the slogan increasingly heralded by the global food sovereignty movement (McKeon, 2015: 83–93).

c The role of academics in supporting social movements' struggles

Another central finding from exploring how LVC is building its model of internationalisation, is how the movement increasingly builds alliances with academics at the local, regional and international level. Chapter 3 of this book showed how a number of academics contribute to the proper functioning of the collaborative engagement in the CSM space by actively supporting the construction of a coherent analysis as well as civil society strategies at the CFS. The need for increased alliance-building has been recognised in a self-evaluation of LVC's engagement in the CFS. This document clearly stated that in order for LVC to engage in the CFS 'undoubtedly demands deployment of more resources, both in time and people, by social movements and their partners' (LVC, 2012a: 1). One way to meet this challenge is to build a relationship with a collective civil society task force that includes academics.

LVC is increasingly attracting academics from a wide range of disciplines seeking to support the analytic capacity of the movement (Claeys, 2013: 21). Some of the academics with close ties to the movement are Jun Borras (one of the founding members of LVC, now professor at the ISS, Den Hague), Annette Aurélie Desmarais (former technical support to LVC and now professor at the University of Manitoba, Canada), Eric Holt-Giménez (Executive Director of Food First, USA), Marc Edelman (City University of New York, USA), Peter Rosset (LVC technical staff member and ex-Food First Director), Philip McMichael (Cornell University, USA), Raj Patel, University of Texas, Austin, USA), Nora McKeon (Terra Nuova, Italy) Josh Brem-Wilson (Coventry University, UK), and Hannah Wittman (Simon Fraser University, Vancouver, Canada). Many of these academics, and others who are not mentioned here, use their positions as researchers to conduct research with and for LVC member organisations at the local, national, regional and global scale.

The contributions from the academic world play a central role in documenting and disseminating the alternative practices and worldviews that emanate from the movement. Building new forms of co-production of knowledge is also central to mobilising counter-power, ideas and concrete proposals to influence the global agenda. Relationship-building with intellectuals is expected to make social movement organisations more efficient in challenging existing power structures, hegemonic discourses and policies (Staggenborg, 2002; Escobar, 1995). In terms of relationships between academics and rural movements such cooperation has been facilitated by the fact that academics themselves are increasingly clustering in networks aiming to support the actions of transnational agrarian movements (McKeon, 2015: 165).

The increased alliance-building between LVC and academics at the global scale is particularly noticeable around food sovereignty. Since LVC first launched the concept in 1996, food sovereignty has rapidly spread beyond peasant movements and has become a vehicle for a much broader global movement of consumers and producers, in both the north, south, urban and rural areas (Claeys, 2015; Bringel, 2015). The broader fieldwork conducted for this book testifies that activist–scholar relations are vigorously being built at the sub-national level. For instance, during a seminar organised by the rural extension of the University de Santa Maria, I witnessed how this reunion benefited from the strong impetus from both youth activists and academics. Strongly inspired by the critical pedagogy of Paulo Freire, Brazilian philosopher and educator, this seminar started with a presentation of his book *Education as the Practice of Freedom* (Freire, 1973). Examples of questions that were raised during the conference were: How can we encourage a new generation to become self-confident to take practical steps to achieving food sovereignty? How can we collaborate with others to put food sovereignty into practical application at all scales?

Whereas relationship-building and constellations of 'action research'[25] between scholars and rural activists mainly take place at the sub-national and regional level, spaces for dialogues are increasingly being established at the global level. The large number of scholars gathered around the Food Sovereignty Colloquia, held at Yale University in late 2013, and early 2014 in The Hague, the Netherlands, corroborated this fact. The latter event alone, organised by the International Institute of Social Studies (ISS), gathered approximately 360 scholars and activists who came together to discuss issues and field experiences in the domain of food sovereignty. Ahead of the conference, 100 selected papers were distributed to the participants as suggested inputs to frame the discussion on food sovereignty. Elisabeth Mpofu, appointed LVC coordinator in 2013, spoke about some of the challenges faced by African farmers in her speech to the audience in the jam-packed conference room: 'How can we be sovereign if we don't have access to land and water? (…) We need to get a better understanding of each other's challenges around the world so that we can better create food sovereignty together' (Field notes, The Hague, 14.01.2014). Another member of LVC emphasised the need to build alliances and to struggle for the need for more 'balanced research':

> 90% of European research funding go to support conventional industrial agriculture, but conventional farming produces only 30% of the food consumed around the world, which is damaging to our planet. We are the majority. We all know now that small-scale farmers feed the world. According to the UN, 70% of the food we consume on the planet comes from small-scale farmers. We need more resources and we need to continue to mobilise many more people around us.
>
> (European Youth member of LVC, interview, The Hague, 14.01.2014)

Although there are still divergent views on the concept and multiple challenges[26] to make it become a reality across scales, food sovereignty has become an exciting research field for scholars from a broad range of disciplines eager to exchange experiences and produce knowledge for sovereignty. The increased gatherings of social movement activists and academics using food sovereignty as a banner of struggles is as an example of the establishment of what Guiso (2000) calls 'collective hermeneutic', namely a space for building convergence around ideas, beliefs, meanings and interpretations.

Finding new ways to forge alliances with academics seems to be central for social movements seeking to build alternative 'facts' directed towards a large-scale transformation for society. Yet, as Chapter 1 showed, while the establishment of valuable links between researchers and social movements have many useful benefits for academics and social activists, such relationships are not unproblematic and must continuously be critically scrutinised (Touraine, 1981; Wieviorka and Calhoun, 2013; Edelman, 2012; Escobar, 2008).

5 Permanent challenges

This chapter ends with a discussion on some of the crucial challenges for LVC as it seeks to increasingly occupy global political space while simultaneously focusing on its grassroots constituencies.

a Everyday distance to the United Nations

How much transnational institutional work, such as norms-building and negotiations in inter-state arenas, actually matters to people's struggle on the ground, remains an issue of debate within members of the movement. Concerns to preserve autonomy, as well as limitations in logistical and human resources seem to remain some of the main apprehensions of rural social movements when interacting with international institutions (Borras, 2004: 18). The geographical distance to UN headquarters in Western capitals contributes to the remoteness that many rural activists express towards the whole UN system.

Peasant activists involved in struggles for social change at the local and national level wonder how time and resources used to engage with international institutions like the UN will produce tangible progress for the peasant effort. For instance, during a Seminar in Rio Grande do Sul, Brazil, I had the chance to debate LVC's engagement in the UN arenas with members of national peasant organisations that belong to LVC.[27] When I presented my fieldwork conducted at the CFS, it led to an intense debate about peasant movements and the ambivalence in relation to the United Nations. It showed that the role of the UN remains a topic of heated debate within the movements. Álvaro Delatorre explained this tension based on his 30 years as a member of the landless movement in Brazil:

> In the eyes of many peasants, international institutions like the United Nations remain dominated by Western values and their interests in defending the status quo. Members are alerted to the fact that participation will lead to the legitimisation of the current neo-liberal order and the agribusiness model. Social movements work for radical social change. Historically, international policy forums like the United Nations have yielded vague results with no binding mechanisms for implementation. We must always ask whether our engagement gives us any new tools to make positive change for people on the ground.
>
> (Álvaro Delatorre, MST, Santa Maria, Brazil, interview, 20.03.2014)

Coordinators of another Brazilian member organisation, the Movimento dos Pequenos Agricultores (MPA), affirmed this tension in an interview with LVC.

The MPA coordinator in Brazil started our meeting explaining that the national organisation was already in Rome, however indirectly, via its membership of LVC. However, as the MPA coordinator added, global policy work at the UN level remains rather abstract for most rural activists engaged in struggles mainly at the national and local level:

> It is not that we do not believe that other members of LVC do not do their best when they engage in UN processes. We believe that our comrades around the world do their job as well as they can. However, many peasant activists find that this level of action is rather distant. These processes are complicated for most peasants to follow. It can involve only few people who can work mainly in English. We do not have much information in Portuguese, which makes it difficult to follow. Dissemination of this kind of work in Portuguese could surely help, but then again, we have only few resources and international work is expensive to engage in.
>
> (MPA Coordinator, Brasilia, Brazil, interview, 30.03.2014)

In the debate, LVC's enhanced engagement with the UN arena and dissemination of information between different levels of engagement remain topical issues, in particular among non-English speakers, who find it difficult to follow the UN work.

This adds to the distance. As the National Coordinator of the Brazilian peasant movement stressed:

> We focus our lobbying mostly on national and regional processes. We find it difficult to participate fully in global meetings because of the language barrier. There is also the other problem with NGOs in global meetings. How can we be sure that they do not occupy these spaces? They are not sincere voices. They have not lived the struggles that social movements have. We should be careful to legitimate spaces if we are not sure we can be there ourselves.
>
> (National Coordinator MPA, LVC, 21.03.2014, interview, Brasilia, Brazil)

National leaders of LVC often emphasise that the global policy work should always be secondary to grassroots mobilisation.

> Our efforts should first and foremost be to support the struggles on the ground and pressure governments at home. When we engage in such a space we must be sure that it does not end up becoming a burden to us at the national level.
>
> (Rita Zanotto, Coordinator of CLOC-LVC, interview, Brasilia, 20.03.2014)

Repeating concerns from LVC members is that the engagement in global policy work risks diverting activists' energy, time and resources away from grassroots activism at the regional, national and local level. The concern is that the time and resources required to participate in laborious UN processes will end up becoming a burden for the national organisations and this seems to be an ever-present tension within the movement.

Such examples of LVC member organisations demanding that delegates of the movement be accountable for their constituents and aware of some of the risks of transnational engagement seems to function as a checks and balance mechanism within the movement. As the MPA member cited above summarises it in a final reflection on rural movements' engagement in global policy processes: 'Our scepticism may be a good one' (National Coordinator, MPA/LVC, interview, Brasilia, Brazil, 21.03.2014).

b The movement as a space for tensions and debates

The strong reflectivity and persistent debate over autonomy within the movement show that LVC does not automatically give up on its political strategy to ensure its survival. This contrasts with the pattern that predicts that social movements engage with others in order to gain access to funding and acceptance from elite allies (Meyer, 1993; Piven and Cloward, 1977; Tarrow, 1998).

Building a coherent global–local and inside–outside strategy for social change remains an assiduous challenge for LVC and may by particularly challenging when LVC moves into complex global policy terrain. It is too simple to reduce the increased engagement at the global level to a 'scale shift' (Bringel, 2015) or a 'zero sum game', in the sense that increased engagement in global policy work means a parallel decline of locality (Albrow, 1996: 10). Nevertheless, peasant delegates do sometimes feel that the laborious work related to global policy work in the CFS can be a sacrifice of time; time that could be used on other types of struggles.

The balance between 'how much time for which kind of work' seems to be a relentless tension in the movement, embodied by members delegated to global work and expected to straddle and translate between global claims-making and grassroots struggles. The limited human and financial resources, as well as the movement's strong focus on a low degree of institutionalisation (cf. Table 2.1), makes the engagement within international structures of, for instance, the UN a particular pertinent challenge. Rosset and Martinez have suggested that the organisational culture of the movement of 'taking the time to consult among members' may explain why LVC has avoided severe internal splits that have plagued other transnational political alliances and coalitions and earlier movements (Rosset and Martinez, 2010: 166; see also Desmarais, 2007). The movement's efforts to remain a space for open dialogue seem to be important for its ability to stay strongly anchored in current debates, manifestations and experiences within the movement.

Fieldwork conducted with LVC members in different regions of the world shows that the relationship between social movements' engagement with (inter)-governmental institutions and grassroots mobilisation continues to provoke tension within the movement. Pleyers contends that tensions related to social movements' strategies and priorities expressed primarily through conflicts and debate among activists themselves must not be mistaken for a diversion or deficiency of the movement: 'On the contrary', the author states, 'keeping these debates alive may in many ways contribute to the movement's dynamics and developments, and the spurring of the movement to adapt to new situations and innovations' (Pleyers, 2010: 193).[28] From this viewpoint, transcending what might seem to be conflicts of interest among different actors and overcoming what seems to be incompatible elements, in order to become complementary aspects, may be both the main vigour and the main challenge of activists seeking to build their force by combining different kinds of political activism.

As one Latin American LVC member reminded the other members of the CSM group in an internal meeting, the engagement in intergovernmental processes is a novel experience for most of the activists seeking to navigate their way to the UN, which to a high degree is a learning process: 'We are new here! We are slowly finding our feet and finding the best strategies on how to form our struggle with others in this kind of space' (Latin American LVC member, Observation, CSM meeting Rome, 10.10.2014).

While effective democratic leadership requires that leaders delegated to global meetings remain responsive to the concerns of the broader member-base, another central aspect of the movement's evolution is that members around the world allow the movement to learn from different kinds of experiences. It means a constant endeavour by LVC to balance the demands of the local, national and regional level with the need for global activism (e.g. Edelman, 1999; Desmarais, 2007).

6 Conclusion

This chapter has discussed some of the main debates and tensions related to LVC as it builds its pattern of internationalisation while engaging with others. By presenting examples of how LVC seeks to keep its autonomy while engaging with others (Borras, 2004; Desmarais, 2007), I show that the movement does not uncritically follow a pattern of aligning with elite allies to secure its survival. Nevertheless, LVC endeavours to influence the global policy agenda and take political leadership in new global policy spaces, with limited human resources. This means that the movement must find ways to build its engagement with others. Forging different forms of alliances is also central for a movement seeking to build its model of internationalisation, political culture and strategies from within its own spaces, as well as in the processes of engaging with others.

How LVC – being simultaneously a global actor and an arena – negotiates its engagement in the global policy sphere in the future is not only crucial for the potential to shape global policy processes on food-related issues. The ongoing political discussions on how to build room for manoeuvring in inter-governmental negotiations and policy arenas is at the heart of the movement's strategy and raises important questions related to the movement's internal functioning. The creation of a coherent strategy by focusing on the development of grassroots constituencies while at the same time forging new types of alliances with the outside world remains both the main driving force and a permanent struggle for the movement. How much 'international structure' a grassroots driven movement like LVC can accept, and how this process affects its internal work and organising, is identified as a key challenge within the movement. With respect to LVC's engagement in the CFS, it seems to be vital that the movement remains a space for internal debate and reflection, as the movement is constantly seeking to finds its feet in this global policy terrain.

Notes

1 Here, Borras refers to the Position Paper 'International Relations and Strategic Alliances' discussed during LVC's 3rd International Conference in Bangalore' (2004:16).
2 The term 'land grabbing' is actively used by LVC and its allies while FAO and other institutions (e.g. World Bank, IFAD) more often refer to 'large-scale land acquisitions'.
3 As LVC is not a legal entity, a process called an 'exchange of letters', carefully expressing interests and terms/conditions of collaboration, is often the way LVC formally engages with others.
4 Despite this episode, after a long exchange of letters in October 2013, a 'New partnership' was built between FAO and LVC:

> FAO will support the effective participation of LVC in political processes at different levels and promote dialogue for designing sustainable local initiatives, projects and emergency interventions. This partnership is based on knowledge sharing, dialogue, policy development and cooperation in normative activities. It will also discuss various issues of mutual interest including those related to land, seeds and agro ecological practices of small scale farmers.
> (FAO, 04.10.2012 http://www.fao.org/news/story/en/item/201824/icode/)

5 I refer to the FAO building which today hosts the CFS and the CSM.
6 For instance, fieldwork conducted in the FAO shows how civil society actors during CFS Sessions often meet with FAO employees as part of their lobby and advocacy work. This was observed, for instance, during the UN *International Year of Family Farming* in 2014.
7 This position was expressed by a US based LVC member in an interview conducted by a former colleague, Lou Pingeot, from the Global Policy Forum, in New York, 16.05.2013.
8 For instance, in a number of Latin American countries during the decades of the 1950s to 1970s, peasants experienced both clientelism and corporatist arrangements where political parties across political spectrums channelled state resources to their corresponding peasant organisations to buy their loyalty (Rosset and Martinez, 2010: 151–152).
9 Other examples of MST activities: beyond their agrarian reform struggles to gaining land titles for families, MST are establishing cooperatives, schools and agro-ecological training courses and mobilising rural people around what Wittman (2009) calls 'agrarian citizenship'.
10 http://www.telesurtv.net/english/news/Landless-Workers-Movement-of-Brazil-MST-Supports-Dilma-Rousseff-20141018-0031.html. This support was strongly contested

in particular after the strong critique of Dilma's mandate during the popular protests in Brazil 2015, where MST and allies in the Frente Brasil Popular demanded a new economic policy for the country. http://www.mst.org.br/2015/09/01/movimentos-e-organizacoes-sociais-promovem-conferencia-nacional-popular.html

11 For instance in Brazil, the country's 1988 constitution opened up the possibility for many experiments with participatory democracy. The most well known is probably the Participatory Budget process that was installed in Porto Alegre in 1989 (Dagnino, 2002; Fung and Wright, 2001).

12 See for instance McKeon (2015: 153).

13 The Food and Agriculture Organization of the United Nations (FAO), The International Fund for Agricultural Development (IFAD) and The World Food Programme (WFP).

14 The point about including civil society from an 'efficiency perspective' in the context of implementation of policies at national level has been communicated by several UN officials during this research, mainly by FAO officials. (Six field research stays at the FAO, Rome during 2013–2015.)

15 It is important to notice that LVC has engaged in alliance-building since its very beginning (see e.g. Desmarais, 2007). However, the focus on alliance-building has been reinforced since LVC's 5th International Conference in Maputo where the political outcome document, among others, emphasised the 'consolidation of a collective identity, providing space for internal debate and alliances with other social movements and carefully selected NGOs' (LVC, 2008: 108).

16 The third sector is often referred to as the massive emergence of NGOs mediating with state and market actors that has led to 'social rights in a neoliberal, patronising and disempowering way' (Gwynne and Key, 2014: 205).

17 In 'Civil Society and Global Democracy: An Assessment' Tallberg and Uhlin (2011) explore accountability among different global civil society actors and identified low mechanisms of internal accountability, in particular among different types of NGOs. For instance, the authors refer to the results of a research group at Syracuse University interviewing leaders of 152 transnational NGOs registered in the United States (Raggo and Schmitz, 2010 cited in Tallberg and Uhlin, 2011: 16). Only 9 per cent of the NGO leaders mentioned accountability to the members' empirical observations while 78 per cent declared that their organisation was accountable to donors.

18 Chapter 2 showed how LVC played a central role in designing the CSM.

19 For instance, I observed during a meeting with a group of French NGOs preparing their participation at the World Social Forum in Tunis how the CSM was criticised as being a 'closed space' with reference to the strong criteria such as the restricted number of delegates they can send to CFS sessions (Observation, civil society meeting, Paris, 25.01 2011). As Pleyers (2010: ch. 2) shows in his book *Alter-globalization*, French NGOs (and intellectuals) have particularly taken a leading role in the global justice movement, leading to a critique from activists, including LVC members. See also Vieira (2011: 217) and Chapter 6 in this book.

20 In particular the financial crisis in 2008 led to a debate about the political and legitimacy crisis of big donor-driven NGOs. This debate emphasised the need for NGOs to review their funding and sustainability models (Ogutu, 2011).

21 Examples of other scholars who have demonstrated the historical challenge of peasant movements working effectively in alliance with NGOs are Edelman (1999) and Desmarais (2007)).

22 What Pleyers referred to as 'Airport activism' during the seminar 'Mouvements sociaux à l'âge global' organised by the Collège d'Etudes Mondiales, Paris, 30.01.2014.

23 The CSM 2010 document specifies the nature of the CSM working groups, where members contribute to the ongoing processes. A member explained that this is an

attempt to prevent certain NGOs from not contributing to the work and using the CSM instead to legitimise their presence in the CFS (CSM member, personal conversation, Rome, 15.10.2013).

24 'Strongly rooted in rural and community movements in all regions, the IPC has combined the political legitimacy and mobilisation capacity of peoples' organisations with the analytic and advocacy skills of NGOs in a mutually reinforcing relationship' (McKeon, 2015: 29).

25 A type of participatory research where the researcher becomes more actively concerned in transforming the research subject/problematic, rather than merely trying to understand it. See Chapter 1.

26 One of the challenges ahead for the broader movement was summarised by Eric Holt-Giménez (executive Director of Food First and ally of LVC) in his closing remarks with the following question: 'Can food sovereignty help us move from a transitional period of multiple crises towards a moment of radical, structural transformation?' (ISS Symposium in Hague, 25.01.2014; speaking notes provided by author).

27 In particular, the Movimento dos Trabalhadores Rurais Sem Terra (MST) and Movimento dos Pequenos Agricultores (MPA).

28 In a similar vein, Fox suggests that a high degree of 'density' or 'cohesion' can best be reached within a movement that allows tensions and differences to exit (Fox, 2000).

Final conclusions and perspectives

Peasants' agency in the global age society

> For more than one hundred years, the great thinkers have predicted that we peasants would disappear. But we have not disappeared. We have resisted wars of extermination, and bad policies designed to drive us off our land and give the food system to the corporations. As peasants we believe that we are on this Earth for a reason. And that reason is to grow food. Food for our families, food for our communities, food for our countries. Healthy food. Food that is grown with respect for the Mother Earth. While we may not have had a high level of formal education, that does not mean we cannot think for ourselves, and organize ourselves into a powerful global movement of resistance. But we are not just resisting, we are also trying to build something new, a better world; with our ideas, and with our actions.
>
> (Elizabeth Mpofu, coordinator of LVC at the Conference on Food Sovereignty in Den Hague, 24.01.2014)

Whereas small-scale producers have often been portrayed as belonging to the past, this book has demonstrated how peasants have planted themselves firmly as important protagonists on the international arena. Against the predictions that peasants would disappear in the processes of modernisation, urbanisation and industrialisation, throughout the last two decades La Vía Campesina has given an unprecedented voice and strength to small-scale farmers at the global level (Desmarais, 2007; Borras and Franco, 2009; Rosset and Martínez, 2010; Menser, 2008; Wittman, 2009).

As a global movement LVC has taken the lead in worldwide protests against neoliberal globalisation and built a global movement from below around an alternative food sovereignty paradigm (Desmarais, 2007; Rosset and Martínez, 2010). The movement also increasingly challenges the global elite from within international institutions, in particular those dealing with issues related to food and agriculture. Exploring LVC as both an actor and an arena (Borras 2004), the analysis in this book has contributed to a better understanding of the dynamics, challenges, debates and tensions arising from a global peasant movement increasingly engaged in multi-scale activities, ranging from local peoples' struggles to global advocacy work. This may be indicative of a recent upsurge of new kinds of agrarian activism (Borras and Franco, 2009; Desmarais, 2007).

While the dominant literature on national social movements leaves less room for innovation and *agency* when social movement activists engage within institutions, this book shows that other models of internationalisation are feasible. In order to demonstrate this, the book has demonstrated how La Vía Campesina carved out a space of engagement within the 2009 reformed Committee on World Food Security. This book has also shown how the global peasant movement is building a much more comprehensive and sophisticated model of internationalisation than predicted by la number of leading scholars studying social movements (Tarrow, 1998; Meyer, 1993; Tilly, 2004; Piven and Cloward, 1977).

In the first part of this book's conclusion, I present how the observations of social movements' comprehensive multi-scale activities compel us to move towards a more dynamic analytical framework. In the second part, I then discuss the role CFS plays as a multi-actor global platform; and in particular how the direct participation of peasants'/peoples' organisations in UN negotiations presents a pioneering example of a more complex, inclusive and democratic foundation for global policy-making. In a concluding perspective, I discuss how mobilised peasants contribute with analyses, practices and production of the global age society (Albrow, 1996; Pleyers, 2010). Lastly, I present two possible future research perspectives related to the building of new foundations for global governance and co-building around knowledge production.

1 Towards a more dynamic approach to analysing global movements

a Social movements – institutional dynamics and limitations

La Vía Campesina as a global movement reveals the agency of a social movement capable of building a more sophisticated model of internationalisation than is usually reflected in most predictions of how social movements' internationalisation will play out (Tilly, 2004; Tarrow, 2005; Keck and Sikkink, 1998).

While dominant theories of social movements' interactions with institutions predict a model of 'routinization, inclusion and cooptation' (Meyer and Tarrow, 1998: 21) with little room for variation, La Vía Campesina's engagement in the CFS proves that social movement activists can build a different pattern. This is particularly evident from the way social movements have managed to design an autonomous interface mechanism for their participation during the 2009 CFS reform process. Contrary to what dominant theories of social movements expect – i.e. that institutionalisation will automatically lead to a process of *homogenisation* (Tilly, 2004; Meyer and Tarrow, 1998) – civil society engagement with the CFS has led to increased *diversity* and participation of more groups of society in order to speak their own voice at the global level.

Despite the challenges arising from representatives of rural constituencies – and other food insecure people – being directly engaged in formal UN processes, this book has revealed the political boldness that permeates social movements. In particular, it demonstrated the institutional originality of social movements and their allies in building an autonomous mechanism to engage with the CFS. This is a concrete example of activists finding ways to strategically combine their claims for autonomy with engagement in a global institution. Besides giving a strong impetus to strategising, networking, and agenda building at the global scale, the CSM is also a space from where different actors can come together to share knowledge and different practice. They can spell out their distinct ideas and engage in profound debate about accountability and representation (Duncan, 2014; McKeon, 2015). This is also an example of how civil society can build common positions based on diversity and creates a space that allows for mutual learning, sharing, reflections and self-assessments.

Bold political consciousness and human reflexivity seem to be central components to avoid some of the traps related to the classical path of institutionalisation (Tarrow, 1998; Tilly, 2004). Although the UN institutional environment exerts an influence on the action repertoire of civil society and social movements' organisations, this book nuances the assertion of other observers who often present institutions as 'iron cages' (Weber, 1958: 181) and point to the danger of movements' engagement with and within institutions (Tarrow, 1998: Tilly, 2004).

Social movements still run the risk of routinisation, elite-formation and co-optation when engaging in an elite institution, yet the dynamics and the human reflexivity registered when rural activists interact with the CFS does not comply with the dominant view of what happens when movements engage in institutionalised politics (Morgan, 2007: 274). Social movements are not only reacting to but also increasingly shaping the processes and institutional dynamics.

b Beyond dichotomies

The observations of mobilised peasants' comprehensive actions and strategies show that rural actors are key actors of contemporary global contestation while simultaneously being rooted in places and communities (Bringel, 2015: 129; Borras and Franco, 2009; see also Desmarais, 2007). While tensions remain, there is not necessarily a contradiction between being an active global movement on one side and a locally 'rooted' movement on the other (Wieviorka, 2014: 3; Bringel, 2015: 129). This observation resonates with Martin Albrow's argument that a 'net increase in globality does not necessarily mean an equivalent decline of locality' (Albrow, 1996: 10: see also Wieviorka and Calhoun, 2013).

A number of scholars have reported how 'globality' is increasingly being shaped in local territories and national contexts and how global struggles may become a focus for local action (Sassen, 2007; Desmarais, 2007; Bringel, 2015; Massey, 1994). While peasants do not move their fields around the world and only a small number of activists engage directly in transnational policy work, data gathered during fieldwork with LVC members in different regions of the world reveal that members of the movement often embody a multi-scale outlook combining the global and local dimensions of their struggle and vision of food sovereignty. Members of the peasant movement, however, often stress that even if food sovereignty is gaining support from some governments who have now incorporated it into national constitutions, its vitality will always come from people's struggle on the ground. As one Latin American peasant farmer expressed it when discussing the role of governments in the struggle for food sovereignty in a civil society meeting in Rome:

> Food sovereignty is way of life that we have cultivated for generations. My grandmother was teaching me about food sovereignty. If some of these governments around the world support us against the dominant model way we can do more. This is why we are here. However, we cannot wait for governments to take action. We must continue to build the world we want in everything we do.
>
> (Latin American LVC member, Rome, 15.10.2014)

While engaging in a formal UN arena may help recognition and ensuring of the right of peoples and communities to define their own food systems, LVC members often emphasise that '99 per cent of the struggle remains on the ground' (European LVC member, CSM meeting evaluation of the CFS annual session, 18.10 2014).

The example of the LVC engagement in a formal UN arena shows that the engagement of social movements at the global policy level cannot be reduced to a 'scale shift' when new political opportunities are constructed (Bringel, 2015: 123). The different kinds of sites and scales in which activists increasingly engage are not necessarily 'either/or' choices. If we wish to explore the possible synergies arising from social movements that engage today in various forms of political activism, we must look beyond the dichotomies such as 'global vs. local', 'radical vs. reformist', and 'outsiders vs. insiders' (Massicotte, 2010: 79; see also Smith and Wiest, 2012: 161). In a similar vein, while supranational structures may offer social movements new political opportunities to address their claims, analysing these configurations from a model of 'political opportunity' structures (Tarrow, 1998; Passy, 2003: 13) does not fully grasp the dynamics of these processes, nor does it help understand the strategic choices, dilemmas, aspirations and agency of social movements (Jasper, 2006).

Exploring how LVC builds its internationalisation must also be seen in the broader changing context of globalisation and ongoing transformations in the rural world. For instance, whereas peasants have struggled with geographic isolation, high costs of communication and transportation and the lack of alternative information, globalisation has lessened these barriers to transnational activism, including for those in the rural areas (Borras and Franco, 2009; Mann, 2014).

If we wish to better understand and explore agency and the possible synergies arising from social movements engaged in multi-scale activities, we need to apply more dynamic analytical frameworks that enable deeper exploration of the patterns of activism that are complex, dynamic, debated, adjusted and negotiated. Finally, all these dynamics call for more multi-scale and multi-site research (Pleyers, 2010: Jasper, 2010a). Conducting this type of research is not only central to our understanding of LVC as a global movement, it is also indispensable to enable us to explore the actors, demands, debates and tensions that constitute and guide the movement (Pleyers, 2010: 13; Touraine, 2001; Wieviorka, 2014).

2 Towards more inclusive global food governance

The interrelated ecology, climate change, and hunger crises have increasingly called into question the dominant global governance system. The crisis has triggered social mobilization around the world of citizens' and people's movements claiming their rights to have a say how the world is shaped. While the state remains the dominant framework of analysis in the social sciences, a state approach does not correspond to today's globalised society and the evolving context of more complex foundations of global policy-making, networked global governance and more comprehensive multi-scale activities (Beck and Cronin, 2006[1]; Bringel and Domingues, 2015; Pleyers, 2015; Wieviorka, 2014; Wieviorka and Calhoun, 2013). What does this mean for the international society and how is the UN as intergovernmental institution seeking to address the complex current challenges? Can global governance be democratized through enhanced mechanisms for inclusion of groups of society whose livelihoods depend on global decisions?

a Linking local struggles to sites of global governance

Several scholars have explored the relationship between international institutions and sectors of civil society seeking to challenge and influence them. Some authors have focused on civil society actors' limited impacts on interstate institutions such as the WTO and the UN processes (Risse, 2002; Steffek et al., 2008; Scholte, 2011). Others, for instance O'Brien et al., demonstrate how global social movements (environmental, labour and women's movements) contesting global governance have actually led towards more 'complex multilateralism' (O'Brien et al., 2000).

The increased visibility of civil society actors in global policy fora affirms that the state is not the only central pillar in the political and social system (Beck and Cronin, 2006; Sassen, 2007). As established civil society groups get better organised around the world, consultation with civil society organisations is largely seen as a benefit that correlates positively with ownership and improvement of the quality of governance (Smith, 2010; Heller, 2013; McKeon, 2009a; O'Brian et al., 2000). Yet, when it comes to actual consultation and inclusion in policy dialogues and processes, the pattern remains often imbalanced.

While peoples' organisations, and particularly those representing the most food insecure, could bring in energy and legitimacy by helping 're-connect' between political authorities and their publics (Scholte, 2011; Dryzek et al., 2011), policy decisions are often taken without inputs from those segments of society who constitute the overwhelming majority of people living in poverty (McKeon, 2009a; Smith and Wiest, 2012).

If we wish to explore how peoples' organisations engage in arenas we need to ask questions such as: What terrain do global institutions offer to peoples' organisations and social movements? How could UN arenas and other international institutions give greater visibility to, and learn from, the energies and experiences of those actors who are simultaneously involved in community struggles and global policy-making.

With the expansion from a purely intergovernmental process to an inclusive global policy forum the CFS can be seen as model of 'alternative governance' rising within the UN (Duncan and Barling, 2012; Duncan, 2015; McKeon, 2015). As Chapter 3 presented, the CFS was reformed with the declared aim to ensure that the voices of all relevant stakeholders, and particularly those most affected by food insecurity, are heard (CFS, 2009). The presentation of the Civil Society Mechanism also showed how it allows for self-organised participation of representatives of peoples' organisations from the sectors of the population most affected by food insecurity (such as peasants, indigenous peoples, fisher-folks) in the policy process leading to decision-making on food and nutrition.

In order to be responsive, effective and inclusive the CFS is not an annual multi-stakeholder event but rather an on-going process where work is undertaken on an on-going basis[2] (CSM, 2010). The UN Committee perceives inclusion of various views and experiences as a prerequisite for effective policy-making from the rationale that broader ownership of a policy-process leads to stronger motivation of citizens to implement its decisions.

However embryonic the actual developments following the 2009 CFS reform may be, the reform presents an example of a more complex, inclusive and democratic governance design where those who may be most affected by the issues under discussion are at the negotiation table with decision-makers. As was discussed in Chapter 3, the CFS multi-stakeholder design presents a rupture of the norm with purely intergovernmental negotiations where civil society engagement (read: NGOs) has been restricted to representatives of accredited international NGOs, occasionally invited as observers to read their statements at the end of sessions when decisions were already made and adopted by governments (McKeon, 2009a: 10). Whereas several challenges remain to achieve the goal of full participation of the direct representatives of those most affected by insecurity (see Chapter 5; Brem-Wilson, 2015) the developments around the CFS reform portray an unprecedented model for affected people's participation and elucidates the steps taken towards more democratic, inclusive and transparent global policy-making (Brem-Wilson, 2015; McKeon, 2015; Duncan, 2014).

This book argues that in order to further explore the dynamics and possible synergies arising from direct participation of peasants – and other citizens – in processes of global decision-making, we cannot restrict our analyses to a focus on how social movements and organisations can build their strategies to engage with international institutions. The structures and norms of institutions and policy arenas can also be changed. Brem-Wilson (2015) has theorised how 'arena adjustment' from the side of the UN could have a positive effect and enable more effective participation of rural and other food insecure constituencies in processes leading to transnational policy-making. Data from background research to this book conducted at the UN Headquarters in New York showed how UN staff and governmental actors are not always familiar with the claims and languages of social movements (Field observations, UNHQ, New York, 2012).

However, increased understanding and recognition of civil society actors' claims for self-organisation and autonomy in their interaction with intergovernmental policy forums seems to be central if the UN and other institutions wish to offer social movements a meaningful space for participation (McKeon, 2009a: 50–120; McKeon and Kalafatic, 2009: 17–18). Recognising that the non-governmental world consists of actors with different mandates and power status also seems to be central if international institutions wish to benefit from the energies and ideas generated by peoples' organisations around the world.

In this regard, Chapter 3 showed how the CFS reform process was remarkable; in particular, the CFS Chair[3] invited civil society actors to engage in an inclusive, informed and political process, self-organising from their autonomous interface-mechanism, separately from the private sector. From the side of social movements, activists may avoid an automatic rejection of 'inside engagement' from preconceived views that this automatically leads to co-option. While internal debates and scepticism are often healthy components of social movements (Pleyers, 2010), the awareness that global institutions and forums differ seems to be important, too.

b Building support for peoples' struggles on the ground

LVC's endeavour to give 'voice' to rural peoples at the global level requires it to engage in global processes where the movement judges there is meaningful space for it to participate. This book has discussed how LVC, in direct response to the neoliberal corporate food regime (McMichael, 2006) seeks to negotiate within global policy spaces like CFS to gain support for their alternative approaches to human rights-based approach in global food governance (McKeon, 2015; Duncan, 2015).

In this regard, the *Voluntary Guidelines on the Responsible Governance of Tenure of Land, Fisheries and Forests in the Context of National Food Security* (VGGTs) requires attention. The VGGTs were developed through an inclusive process involving a series of consultations and negotiations with social movements, farmer associations, other CSOs, representatives of international institutions, the private sector and academia. The inclusive process led to the development of the first globally negotiated guidelines on responsible governance of tenure

(FAO, 2012b). As discussed in Chapter 2, the VGGTs are situated in a context of increased land grabbing and decades of struggles for peoples' access to and control over natural resources (McKeon, 2015). Whereas disagreements and controversies naturally exist within a dynamic and diverse movement,[4] according to a statement from LVC, the Tenure Guidelines present advances in land governance and are seen as 'a major step towards a human rights based governance of natural resources' (LVC, 2015).

Since the VGGTs were endorsed by governments in the CFS in 2012, LVC has strengthened its global and regional networks and organised workshops for its members to raise awareness about and how to use the VGGTs to serve peoples' struggles to defend human rights and ensure justice in the process of governance of natural resources. In order to make the technical global guidelines more accessible for people LVC, together with larger food sovereignty movements, is engaged in the dissemination and 'popularizing' of the VGGTs by adapting and translating them into a more relevant peoples' language in the *People's Manual on the Tenure Guidelines.*[5] The objective is that this manual – combined with support and training – can build local-level advocacy by marginalized groups, who can get to understand the VGGTs and deploy them in their struggles (Ortega-Espés et al., 2016).

However, as this book has demonstrated (Chapters 6 and 7), to transform global negotiated documents to serve people's national struggles requires substantial efforts from all actors involved. For the representatives of LVC, building effective linkages between global policy work and local struggles is further complicated by the fact that the movement is at once an actor and arena of debate among different national and sub-national peasants' and farmers' groups (Borras and Franco, 2009: 11) with interests in and resources to engage in different types of work. It remains for most human beings a challenging task to do the balancing act of translating building links and synergies between global policy work and local peasants' struggles. These challenges, however, do not prevent peasants in seeking to overcome them.

> Global policy documents and outcomes remain abstract unless people take them home and use them in their local communities. The challenge is to educate our own members about their rights and how to use the guidelines. Citizens around the world face severe consequences when promoting rights at the national level. We need mechanisms to secure people's access to land and livelihood and safeguard people against dispossession and environmental damage.
>
> (African LVC member, interview, CFS Annual, Rome, 14.10.2013)

As the member of LVC expresses here, for a peasant movement like LVC, lobbying and advocacy remain only *means* through which they can win support for their struggles and access to and control over land and natural resources. LVC and other rural activists with knowledge of the CFS and its products often refer to the recommendation of the establishment of spaces for dialogue between states and food producers.[6] In particular, the aspiration is that peasant organisations and other civil society actors may use the VGGTs to build open spaces for active, meaningful and effective civil engagement in the formulation of laws, policies and programmes (European, LVC member, interview, CFS, Rome, 15.10.2015). Equally important for civil society actors is the use of the CFS arena to hold policy-makers responsible for their actions towards realisation of the right to adequate food at global, regional and national levels.

Hence, whereas the fields and communities remain the main sites for most peasants' everyday struggles for food sovereignty, the engagement in a global policy space like the CFS may be one component in the transition towards a human rights based foundation for global food governance anchored in more just, resilient, and sustainable food systems (McKeon, 2015; Duncan, 2015).

c Peasants envisioning a path for global society

A central component of conceptualising social movements as a heuristic tool (see Chapter 1) is to accept the premise that social movements can help us to better understand the dimensions of society, life and social transformations (Touraine, 1981: 29; Wieviorka, 2013; Melucci, 1996: 25). This book has shown how mobilised peasants move from the plough to the negotiation halls of the UN and struggle against the dominant industrial model of agriculture, which has devalued small-scale farming, local cultural practices and knowledge (Marglin, 1996; Desmarais, 2007; McMichael, 2010).

In contrast to the episteme that has demanded peasants to 'modernise' as quickly as possible in order to 'catch-up' with Western technologies, thinking and practices (Desmarais, 2007) LVC is an example of a social movement showing the possibility of an ontological alternative (McMichael, 2010). Against the narrative that rural actors are 'backwards' or 'obsolete', mobilised peasants today engage in sophisticated global networks and complex scientific debates related to food, climate, trade and investments. By presenting food sovereignty as the broader vision, mobilised peasants project themselves as key agents of large-scale transformations towards a model of food production and consumption that is better suited to responding to complex global challenges than the dominant model of agriculture (Mann, 2014; Desmarais, 2007).

As a growing movement, LVC is establishing its legitimate presence in the international arena and challenges the, for many, unthinkable agency of peasants in a modern world. The movement shows how peasants, indigenous peoples and other historically marginalised groups of society, are today presenting themselves as the solution to feeding the planet, protecting the environment and limiting global warming. With the range of multi-scale activities and strong consciousness of the need to connect local and global struggles, the peasant movement contributes to the evolving context of what Martin Albrow calls the 'global age society' characterized by an increased consciousness of the interdependence at the scale of humanity and the finitude of the planet (Albrow, 2015; see also Pleyers, 2015: 3).

This observation resonates with the argument that LVC's engagement in a global policy arena in CFS cannot be reduced to a question of influence, voice, or recognition per se. When peasant representatives of LVC engage in global policy deliberations and negotiations to build democratic, transparent and just food systems they do not only present this as a struggle for their own constituencies. LVC presents this as a struggle for food sovereignty for all peoples of the world. It demonstrates that other paths for society are possible – and already visible in the existing alternatives around the world.

As was shown in Chapter 4, when social movements challenge 'facts' of the dominant agricultural model social activists engage in a battle over which kind of knowledge and practices count as 'evidence' for global policy-making, and this is part of a broader struggle over shaping society, what Touraine calls the 'historicity of society' (1973: 354). Whichever direction the ongoing struggle over the major orientation of society takes, today peasants have planted themselves firmly in the global policy arena with people-driven alternatives to the dominant global food regime. These coherent alternatives, emerging from people's everyday struggles around the world, are gaining increased resonance in the global age society (Albrow, 1996) that encourages citizens to come together and contribute to the building of a fairer, sustainable and more democratic co-existence of human beings in a global world (Pleyers, 2010).

3 Building convergences and synergies

The reconfigurations arising from social movements' engagement in a global policy arena open a panorama for further research perspectives. Food sovereignty activists such as the global social movement La Vía Campesina and the International Planning Committee for Food Sovereignty (IPC) are increasingly seeking to shape the direction of food and agricultural global governance. This book has shown how these efforts have more recently been extended to social movements' participation in the 2009 reformed CFS via the CSM. Another example of LVC taking a lead role in an intergovernmental arena is the movement's engagement in the UN Human Rights Council and the drafting of the *Declaration on the Rights of Peasants and Other People Working in Rural Areas*. The aim of this process is to build support for the defence and protection of the rights of peasants and other rural actors who are often exposed to disproportionate discrimination, threats and violence (Claeys, 2015; Monsalve Suárez, 2013).

a. Rural movements and international institutions

As representatives of peasants' movements, indigenous peoples, pastoralists and other groups that are increasingly organising themselves as peoples' organisations are finding their way to the negotiating table within international institutions, further research could explore and compare some of these experiences and lessons learned. For instance, what are the main challenges ahead to build inclusive processes for people to have a meaningful participation in decisions that impact their everyday lives? Which arenas truly leave space for continuous social movement and alternative building, how and why? How do the diverging mandates of these institutional sites affect civil society engagement and shape the overall discourse? This last question is a research agenda being pursued by Carolin Anthes (Anthes, forthcoming).

Comparing the experiences in different policy arenas would help to better understand the opportunities for and obstacles to people having a meaningful voice in global policy-making – but also how these global processes are shaped and can feed into peoples' struggles on the ground. Social movements engaging in the CFS – as well as in other policy arenas where they have carved out a space for participation, such as in the UNHRC – constantly need to analyse how and whether their engagement in policy arenas benefits national and local struggles. How global products are articulated with national level policy-making and to what extent civil society can utilise global governance instruments in a country context and local struggles is an important research agenda for the future. In this regard, the research agenda of Katie Whiddon, who is interrogating the potential impact of the Tenure Guidelines on local struggles for food sovereignty – and the factors that condition it – will be extremely valuable (Whiddon, forthcoming).

Further research on the relationship between social movements and international arenas will contribute to understanding global movements' struggles and the transformative potential of social movements today, actively forcing both horizontal and vertical networks to influence global policy-making (Borras and Franco, 2009; McKeon, 2015). More research on social movements' participation in international governance arenas will also be important to go beyond 'top-top' and 'bottom-up' towards a more complex understanding of the evolving context of global governance (McKeon, 2015; Bringel, 2015). Future research will not only depend on how researchers gain access to participate in these policy arenas, but also, more importantly, to which degree researchers engage in multi-site and multi-scale research to document how the global–local dynamics play out.

b. Co-building around knowledge production

The CFS as a multi-actor platform presents an example of how the expression of various views and experiences in the policy processes may provide valuable and essential expertise in the identification of problems and solutions (CFS, 2010). This opens up another research perspective to better explore such new constellations of knowledge production. For researchers seeking to analyse possible co-building around knowledge the High Level Panel of Experts (HLPE) requires special attention. As explained in Chapter 2, a central pillar of reforming the CFS in 2009 was the establishment of an inclusive science panel as an interface to its policy processes. This led to a unique institutional structure, where a scientific body complements the political one. The declared goal of the HLPE is to 'help to create synergies between world class academic/scientific knowledge, field experience, knowledge from social actors and practical application' (CFS, 2009: 36; McKeon, 2015: 85).

As a scientific panel integrating knowledge from various scientific sources to strengthen the foundation for global policy-making on food and nutrition, the HLPE can be seen as an example of a transdisciplinary research panel at the global scale. As a reflective, integrative, and inclusive scientific panel directly linked to a UN body, the HLPE opens up a window of opportunity for epistemological discussions and further research perspectives on how to build new constellations for knowledge production and decision-making (Lang et al., 2012).

The increased awareness of the benefits from greater involvement of actors from outside academia into the research and policy processes, in order to integrate the best available knowledge and create ownership for problems and solutions, has led to an increased interest in transdisciplinarity research and the broader context of integrative research methods, where participatory research, action research and others are also located (Rosendahl et al., 2015; Lang et al., 2012).

One of the foundations of these epistemological perspectives is the acknowledgment of the value of non-scientific knowledge. The aim is to build bridges between non-scientific (practical, operational, traditional, indigenous, etc.) and scientific knowledge, and improve our general understanding with a view to intervene and transform society towards sustainable goals. A recent example of this 'transformational scientific field' aiming to better integrate science and society and comprising perspectives from multiple disciplines and different types of knowledge is the International Panel of Experts on Sustainable Food Systems (IPES-Food).[7]

The emergent field of transformative research on interlinkages and synergy building between different sources of knowledge from the world of social movements, academia and international institutions all seem to be important for activists, researchers and the global policy elite alike.

With empirical data collected in a constant back-and-forth between academia, the UN and social movement arenas, the hope is that this book can help to better understand the challenges and potentials for peasant movements to build new processes of internationalisation, institutionalisation and participation.

Notes

1 In his argument against 'methodological nationalism', Ulrich Beck analyses modern society arguing that we cannot limit our lenses to territorially limited nation states and suggests that we should move towards a 'cosmopolitan methodology' (CADIS-seminar, EHESS, Paris, 11.12.2013). Beck argues that we need to replace the linearity assumption of the 'either–or' of national axiomatic by more dialectic 'both–and' postulates: globalisation and regionalisation, linage and fragmentation, centralisation and decentralisation, are dynamics that belong together as two sides of the same coin (Beck 2000: 45).

2 CFS is organised around a number of thematic Open Ended Working Groups that publish their agendas and meetings notes. http://www.fao.org/cfs/workingspace/cfs-ws-home/en/
3 The then CFS Bureau Chair, the Argentinean Permanent Representative to FAO, Maria del Carmel Squeff during the CFS reform process is seen by civil society actors and observers as having played a key role as 'champion' helping to get the political by-in from other governments, in particular Brazil and a number of European and African countries in the CFS-reform process (McKeon, 2015: 106).
4 Discussions with members of LVC during meetings and seminars at the national level revealed one central axis of the debate related to social movements (dis)engagement with global processes such as the VGGT: whether global guidelines or principles adopted by governments in UN intergovernmental fora will support people's struggle on the ground or end of existing land relations at the cost of a transformative visions of agrarian reform (Fieldwork, Santa Maria, Rio Grande de Sul, March 2014).
5 http://www.foodsovereignty.org/peoplesmanual/
6 Paragraph 26.2 of the *Voluntary Guidelines on the Responsible Governance of Tenure of Land, Fisheries and Forests in the Context of National Food Security*: http://www.fao.org/docrep/016/i2801e/i2801e.pdf
7 The IPES-Food was launched in 2015 and describes itself as 'A new panel guided by new ways of thinking about research, sustainability, and food systems': http://www.ipes-food.org/first-report-from-ipes-food-who-shapes-food-systems-and-who-has-a-say-in-how-they-are-reformed

Appendix A

Field research for this research has been conducted in five regions (Europe, North America, South East Asia, Indonesia, West Africa and Latin America (see Figure A1. I have observed and interviewed peasant actors in different local, national and international gatherings.

1. *Italy, Rome.* Researcher in the CFS Open-Ended Working Group of Principles for Responsible Agricultural Investment (CFS-rai) as well as the CSM Open-Ended Working Group of Principles for Responsible Agricultural Investment.

   April–May 2013: Observations of civil society preparations meeting for the Open-Ended Working Group on Responsible Agricultural Investment (CFS-rai).

   5–6 October, 2013: The 4th Annual Forum of the International Food Security & Nutrition Civil Society Mechanism (CSM).

   7–11 October 2013. Rome, Italy. The 40th Plenary Session of CFS.

   19–24 May 2014. Rome, Italy. CFS-rai Global Meeting; 1st round of CFS-rai negotiations.

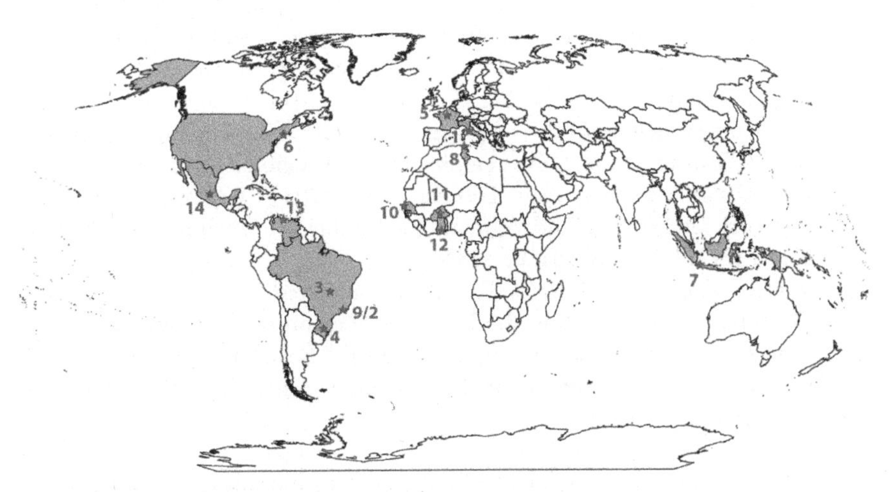

Figure A.1 Map of main field sites (2005–2015)

4–8 August 2014. CFS-rai Global Meeting. 2nd round of CFS-rai negotiations.
11–12 October 2014. The 5thAnnual Forum of the International Food Security & Nutrition Civil Society Mechanism (CSM).
13–18 October. 2014. The 41th Plenary Session of the CFS.
10–11 October, 2015. The 6thAnnual Forum of the International Food Security & Nutrition Civil Society Mechanism (CSM).
12–17 October. 2015. The 42th Plenary Session of the CFS

2. *Brasilía/Rio de Janeiro, Brazil.* 1–30 September 2014: Research with LVC members and academic network at CPDA/Universidade Federal do Rio de Janeiro/Brasilia Brazil, Interviews with LVC/CLOC secretariat and with The Minister of Agrarian Development (MDA), Pepe Vargas.
3. *Vale de Jequitinhonha, Minas Gerais, Brazil.* July–August, 2013. Field stay with Movimento dos Trabalhadores Sem Terra (MST)
4. *Rio Grande De Sul, Brazil.* 20–25 March 2014. Participation in and presentations at Seminars of the Rural Extension at the University of Santa. Field research with members of MST and Movimento dos Pequenos Agricultores (MPA) in Santa Cruz, Rio Grande de Sul. Brazil.
5. *Nice, France.* 'People's Forum' Counter-summit against G8. 1–10 November, 2011.
6. *New York, USA.* UN Headquarters. August–December 2012 and again in October–December 2012. Conducting research as an associate (2008) and Visiting Fellow (2012) at the UN Headquarters through the Global Policy Forum, an NGO monitoring the work of the United Nations.
7. *Jakarta, Indonesia.* La Vía Campesina's 6th International Conference + Field Trip to Sukabumi/Sinarjaya. West Java, Jakarta, Indonesia, 11–17 June 2013.
8. *Tunis, Tunisia.* World Social Forum 2013. Tunisia, Tunis 26–30 March,
9. *Rio de Janeiro, Brazil.* 'Rio+20 People's Summit', 15–23 June 2012.
10. *Dakar/Saint Louis, Senegal.* Participation in the World Social Forum in Dakar, Senegal, 6–11 February 2011. Participating in the drafting of the collective appeal against land grabbing and field research conducted with the Conseil national de concertation et de coopération des ruraux (CNCR) in the Saint Louis region (northern region of Senegal).
11. *Quagadougou, Burkina Faso.* February–April 2011. Research within the Network of Farmers' and Agricultural Producers' Organisations of West Africa (POPPA) and Africa LandNet.
12. *Tamale, Ghana.* April 2010. Research conducted on land grabbing and the extractive industries in the Northern Region of the Country.
13. *Caracas, Venezuela,* September 2006. World Social Forum and research with the social movements of the Bolivarian Alliance for the Peoples of Our America (ALBA).
14. *Oventic, Chiapas, Mexico.* November 2005 to January 2006. Field research with the Zapatista movement in indigenous communities of Chiapas

Appendix B

For this book I interviewed 91 respondents and conducted over 140 semi-directed interviews. The following tables present an overview over the categories and regions that these respondents belong to.

Category	
LVC members/LVC technical support	*33*
Other representatives of civil society	22
UN diplomats/governmental actors	23
UN staff	16
Financial institutions and private sectors representatives	3
Professors/academics	16

Region	
EU	*31*
North America	9
Latin America	24
Asia	10
Africa	12

Appendix C

a Declaration of Nyéléni

27 February 2007

Nyéléni Village, Sélingué, Mali

We, more than 500 representatives from more than 80 countries, of organizations of peasants/family farmers, artisanal fisher-folk, indigenous peoples, landless peoples, rural workers, migrants, pastoralists, forest communities, women, youth, consumers,

What are we fighting for?

A world where...

...all peoples, nations and states are able to determine their own food producing systems and policies that provide every one of us with good quality, adequate, affordable, healthy, and culturally appropriate food;

...recognition and respect of women's roles and rights in food production, and representation of women in all decision making bodies;

...all peoples in each of our countries are able to live with dignity, earn a living wage for their labour and have the opportunity to remain in their homes;

...where food sovereignty is considered a basic human right, recognised and implemented by communities, peoples, states and international bodies;

...we are able to conserve and rehabilitate rural environments, fish stocks, landscapes and food traditions based on ecologically sustainable management of land, soils, water, seas, seeds, livestock and other biodiversity;

...we value, recognize and respect our diversity of traditional knowledge, food, language and culture, and the way we organise and express ourselves;

...there is genuine and integral agrarian reform that guarantees peasants full rights to land, defends and recovers the territories of indigenous peoples, ensures fishing communities' access and control over their fishing areas and eco-systems, honours access and control over pastoral lands and migratory routes, assures decent jobs with fair remuneration and labour rights for all, and a future for young people in the countryside;

...where agrarian reform revitalises inter-dependence between producers and consumers, ensures community survival, social and economic justice and ecological sustainability, and respect for local autonomy and governance with equal rights for women and men

...where it guarantees the right to territory and self- determination for our peoples;

...where we share our lands and territories peacefully and fairly among our peoples, be we peasants, indigenous peoples, artisanal fishers, pastoralists, or others;

...in the case of natural and human-created disasters and conflict-recovery situations, food sovereignty acts as a kind of "insurance" that strengthens local recovery efforts and mitigates negative impacts

...where we remember that affected communities are not helpless, and where strong local organization for self-help is the key to recovery;

...where peoples' power to make decisions about their material, natural and spiritual heritage are defended;

...where all peoples have the right to defend their territories from the actions of transnational corporations;

What are we fighting against?

Imperialism, neo-liberalism, neo-colonialism and patriarchy, and all systems that impoverish life, resources and eco-systems, and the agents that promote the above such as international financial institutions, the World Trade Organization, free trade agreements, transnational corporations, and governments that are antagonistic to their peoples;

The dumping of food at prices below the cost of production in the global economy;

The domination of our food and food producing systems by corporations that place profits before people, health and the environment;

Technologies and practices that undercut our future food producing capacities, damage the environment and put our health at risk. Those include transgenic crops and animals, terminator technology, industrial aquaculture and destructive fishing practices, the so-called white revolution of industrial dairy practices, the so-called 'old' and 'new' Green Revolutions, and the 'Green Deserts' of industrial bio-fuel monocultures and other plantations;

The privatisation and commodification of food, basic and public services, knowledge, land, water, seeds, livestock and our natural heritage;

Development projects/models and extractive industry that displace people and destroy our environments and natural heritage;

Wars, conflicts, occupations, economic blockades, famines, forced displacement of people and confiscation of their land, and all forces and governments that cause and support them; post disaster and conflict reconstruction programmes that destroy our environments and capacities;

The criminalization of all those who struggle to protect and defend our rights;

Food aid that disguises dumping, introduces GMOs into local environments and food systems and creates new colonialism patterns;

The internationalisation and globalisation of paternalistic and patriarchal values that marginalise women, diverse agricultural, indigenous, pastoral and fisher communities around the world;

What can and will we do about it?

Just as we are working with the local community in Sélingué to create a meeting space at Nyéléni, we are committed to building our collective movement for food sovereignty by forging alliances, supporting each others' struggles and extending our solidarity, strengths, and creativity to peoples all over the world who are committed to food sovereignty. Every struggle, in any part of the world for food sovereignty, is our struggle.

We have arrived at a number of collective actions to share our vision of food sovereignty with all peoples of this world, which are elaborated in our synthesis document. We will implement these actions in our respective local areas and regions, in our own movements and jointly in solidarity with other movements. We will share our vision and action agenda for food sovereignty with others who are not able to be with us here in Nyéléni so that the spirit of Nyéléni permeates across the world and becomes a powerful force to make food sovereignty a reality for peoples all over the world.

Finally, we give our unconditional and unwavering support to the peasant movements of Mali and ROPPA in their demands that food sovereignty become a reality in Mali and by extension in all of Africa.

Now is the time for food sovereignty!

b The seven principles for responsible agricultural investment that respects rights, livelihoods and resources

A discussion note prepared by FAO, IFAD, UNCTAD and the World Bank Group to contribute to an ongoing global dialogue. January 25, 2010*

1 Land and resource rights: Existing rights to land and natural resources are recognized and respected.
2 Food security: Investments do not jeopardize food security, but rather strengthen it.
3 Transparency, good governance and enabling environment: Processes for accessing land and making associated investments are transparent, monitored, and ensure accountability.
4 Consultation and participation: Those materially affected are consulted and agreements from consultations are recorded and enforced.

5 Economic viability and responsible agro-enterprise investing: Projects are viable in every sense, respect the rule of law, reflect industry best practice, and result in durable shared value.
6 Social sustainability: Investments generate desirable social and distributional impacts and do not increase vulnerability.
7 Environmental sustainability: Environmental impacts are quantified and measures taken to encourage sustainable resource use, while minimizing and mitigating the negative impact.

*Each of the seven RAI Principles is presented with a section explaining 'Why it matters' and 'What can be done' (FAO et al., 2010).

c The Principles for Responsible Investment in Agriculture and Food Systems were approved the 41st Session of CFS on 15 October 2014.

1 Contribute to food security and nutrition: sustainability and quality of production, equity in income generated, transparency and efficiency of markets, better conditions for food utilization
2 Contribute to sustainable and inclusive economic development and the eradication of poverty: employment generation, respect of principles and rights defined in the ILO core conventions, enforceable and fair contracts, partnerships and cooperation, sustainable consumption
3 Foster gender equality and women's empowerment: non discrimination, equity, access to land, participation
4 Engage and empower youth: access to land, training
5 Respect tenure of land, fisheries, forests and access to water: respect of legitimate tenure rights
6 Conserve and sustainably manage natural resources, increase resilience, and reduce disaster risks: air, land, soil, water, forests, biodiversity and genetic resources, wastes and losses, resilience, climate change, good technological practices.
7 Respect cultural heritage and traditional knowledge, and support diversity and innovation: traditional knowledge, skills and practices, role of indigenous peoples and local rural communities in agriculture and food systems, contributions of farmers to seed selection
8 Promote safe and healthy agriculture and food systems: food safety, animal welfare, plant health, inputs and the environment
9 Incorporate inclusive and transparent governance structures, processes, and grievance mechanisms: rule and application of law, sharing of information, participation, mediation, rights.
10 Assess and address impacts and promote accountability: indicators, independent evaluations, appropriate and effective remedial and/or compensatory actions.

(FAO, 2014b. For full presentation see: http://www.fao.org/3/a-ml291e.pdf)

d CFS in Rome: The majority of governments remain blind to the challenges of global food security

La Vía Campesina press release

Rome 15 October 2014

The delegation of La Vía Campesina, gathered in Rome for the 41st session of the Committee on World Food Security (CFS), recognizes the CFS as the major international forum for debate and decision-making on agricultural and food issues. LVC urges governments to take urgent action in favor of peasant and indigenous agriculture, which is the only model capable of feeding the world. On the occasion of World Food Day, we restate our commitment to struggle for Food Sovereignty as a solution to the multiple crises affecting our societies. We reaffirm our commitment to the recognition and enforcement of peasant rights.

The celebration of the 10th anniversary of the Guidelines on the Right to Food has shown a huge gap between rights and their priority, respect, and application in reality. In this sense, LVC expressed deep disappointment with the lack of commitment to the application of the Guidelines.

Kannayian Subramaniam, a farmer from the state of Tamil Nadu in India denounced the attacks in the WTO to the food reserves created in India: "Public stock holding is vital to the food and nutrition security of any country. It is one of the main weapons that we have against food price volatility. Any trade measure that comes in the way of countries assisting the poorest and most marginalized people is unacceptable to us. The principle of coherence of human rights overrides any trade negotiation or agreement that comes in the way of food security of our constituent groups." LVC confirms that it is essential to discuss market rules within the CFS.

The adoption of Principles for responsible investment in agriculture (rai) is not sufficient to guarantee the rights of peasant communities, landless people and agricultural workers. It is positive that the primary role of peasants in investment in agriculture is recognized prior to the recognition of the role of the corporate sector. However, the rai do not give clear and strong guidance in the interest of the small-scale producers.

The guidelines do not contain sufficient safeguards to stop land grabbing and other destructive actions by private capital and complicit governments. No real progress in promoting the creation of decent work, workers' rights, and in the fight against discrimination of women was made.

As mentioned by Javier Sanchez, a peasant farmer from Aragón: "We need public policies in favor of food sovereignty, promoting agroecology, local markets, the empowerment of women, access to the profession for young people and access to and control over land, forests, water and seeds." La Vía Campesina expresses the need for the CFS to take a greater role in the design of agricultural and international food policies. We recognize the progress made since its reform and are committed to further promote policies that address the needs of the most excluded populations. LVC urges the CFS to launch processes to develop policies that support stable markets and agroecological agriculture, which are respectful of human and peasant rights. These policies must also contribute to stop climate change, ensure access to resources such as seeds and water and put the public interest before private interests.

e The Jakarta Call

Call of the VI Conference of La Vía Campesina – Egidio Brunetto (9–13 June 2013)

Jakarta, 12 June 2013

We, La Vía Campesina, call rural and urban organizations and social movements to transform and build a new society based on food sovereignty and justice. We gather here strengthened by the spirit of our friends and leaders and all those whose courage and commitment to our struggle inspires us. La Vía Campesina, an international peasant movement gathers more than 200 million peasants, small-scale producers, landless, women, youth, indigenous, migrants, and farm and food workers, from 183 organizations and 88 countries. We are here in Asia home to the majority of the world's peasants to celebrate our first two decades of struggles.

Coming together in Mons (Belgium) in 1993 and articulating our radical vision of food sovereignty in Tlaxcala (Mexico) in 1996, we have succeeded in repositioning peasant and family farmers (men and women) as a central social actor in the processes of resisting the neo-liberal trade agenda and constructing alternatives. As people of the land we are vital actors not only in the construction of a distinct agricultural model, but also in building a fair, diverse and egalitarian world. We feed humanity and care for nature. Future generations depend on us to protect the earth.

Today, more than ever another world is necessary. The destruction of our world, through overexploitation and dispossession of people and the appropriation of natural resources is resulting in the current climate crisis and deep inequalities which endanger human kind and life itself. La Vía Campesina says a resounding NO to this corporate-driven destruction.

We are building new relationships between human beings and nature based on solidarity, cooperation and complementarity. At the heart of our struggle is an ethic of life. La Vía Campesina is committed to giving visibility to all of the local struggles around the world, ensuring that these are understood from international perspectives and integrated into a global movement for food sovereignty, social change and self-determination for the peoples of the world.

We call on our organizations, allies, friends, and all those committed to a better future to reject the 'green economy' and build food sovereignty.

OUR WAY FORWARD

FOOD SOVEREIGNTY NOW—TRANSFORMING OUR WORLD

Food sovereignty is a key part of the fight for social justice bringing together many sectors from the countryside and the city. Food sovereignty is the fundamental right of all peoples, nations and states to control food and agricultural systems and policies, ensuring every one has adequate, affordable, nutritious and culturally appropriate food. This requires the right to define and control our methods of production, transformation, distribution both at the local and international levels.

During the last two decades our vision of food sovereignty has inspired a generation of activists engaged in social change. Our vision for our world encompasses an agricultural revolution as well as socio-economic and political transformation. Food sovereignty articulates the crucial importance of local and sustainable production, respect for human rights, fair food and agricultural prices, fair trade between countries, and the safeguarding of commons against privatization.

Today, we are facing a major crisis in our history, which is systemic. Food, labour, energy, economic, climate, ecological, ethical, social, political and institutional systems are collapsing in many parts of the world. The growing energy crisis in a context of fossil fuel depletion is being addressed with false solutions ranging from agrofuels to nuclear energy which are among the greatest threats to life on earth.

We reject capitalism, which is currently characterized by aggressive flows of financial and speculative capital into industrial agriculture, land and nature. This is generating huge land grabs and a brutal displacement of people from their land, destroying communities, cultures and ecosystems. It creates masses of economic migrants, climate refugees and unemployed, increasing existing inequalities.

Transnational corporations, in complicity with governments and international institutions, are imposing under the pretext of green economy GM monocultures, mega mining, dams and fracking projects, large tree and bio-fuel plantations, or the privatisation of our seas, rivers, lakes and forests. Food sovereignty wrests control over our commons back into the hands of the people.

AGROECOLOGY IS OUR OPTION FOR TODAY AND THE FUTURE

Peasant agriculture, artisanal fisheries and herding remain the source of most of our food. Peasant agroecology is a social and ecological system encompassing a great diversity of technologies and practices that are culturally and geographically rooted. It removes dependencies on agro-toxins, rejects confined industrial animal production, uses renewable energies, and guarantees healthy food. It enhances dignity, honours traditional knowledge and restores the health and integrity of the land. Food production in the future must be based on a growing number of people producing food in more resilient and diverse ways.

Agroecology defends biodiversity, cools down the planet and protects our soils. Our agricultural model not only can feed all of humanity but is also the way to stop the advance of the climate crisis through local production in harmony with our forests and waterways, enhancing diversity and returning organic matter to natural cycles.

SOCIAL AND CLIMATE JUSTICE, AND SOLIDARITY

As we build upon our geographical and cultural diversity our growing food sovereignty movement is reinforced by integrating social justice and equality. Practicing solidarity over competitiveness, rejecting patriarchy, racism, colonialism and imperialism, we struggle for participatory and democratic societies, free from the exploitation of children, women, men and nature.

We demand climate justice now. Those who are suffering most are not those who generate climatic and environmental chaos. The drivers of capitalist growth through the false solutions of the green economy are worsening the situation. Therefore, ecological and climate debt must be rectified. We demand the immediate stop to carbon market mechanisms, geo-engineering, REDD, and agro-fuels.

We will keep fighting permanently against transnational corporations, by among other actions, boycotting their products and refusing to cooperate with their exploitative practices. Free trade and investment agreements have created conditions of extreme vulnerabilities and injustices for millions. The implementation of these agreements results in violence, further militarization and the criminalization of resistance. Another tragic outcome is the massive movement of peoples migrating to low-paid, insecure and unsafe jobs rife with human rights violations and discrimination. La Vía Campesina has succeeded in putting the rights of peasants on the agenda of the Human Rights Council and we call on national governments to realize these rights. Our struggle for human rights is at the heart of international solidarity and includes the rights and social protections of migrant farm and food workers.

A WORLD WITHOUT VIOLENCE AND DISCRIMINATION AGAINST WOMEN

Our struggle is to build a society based on justice, equality and peace. We demand respect for all women' rights. In rejecting capitalism, patriarchy, xenophobia, homophobia and discrimination based on race and ethnicity, we reaffirm our commitment to the total equality of women and men. This demands the end to all forms of violence against women, domestic, social and institutional in both rural and urban areas. Our Campaign against Violence towards Women is at the heart of our struggles.

PEACE AND DEMILITARIZATION

There is an increase in conflicts and wars over appropriation, proliferation of military bases and criminalization of resistance. This violence is intrinsic to a deadly capitalist system based on domination, exploitation and pillage. We are committed to respect, dignity and peace.

We grieve and honour the hundreds of peasants who have been threatened, persecuted, incarcerated, and even killed in their struggles. We demand accountability and punishment for those who violate human rights and the rights of nature. We also demand the immediate release of political prisoners.

LAND AND TERRITORIES

We demand a Comprehensive Agrarian Reform. This means ensuring full rights over land, recognizing indigenous peoples' legal rights to their territories, guaranteeing fishing communities' access and control of fishing areas and ecosystems, and recognising pastoral migratory routes. Only such reform ensures a future for young rural peoples.

A Comprehensive Agrarian Reform also includes a massive distribution of land as well as livelihood and productive resources to ensure permanent access to land for youth, women, the unemployed, the landless, displaced, and all those willing to engage in small-scale agroecological food production. Land is not a commodity. Existing laws and regulations need to be reinforced, while new ones are needed to protect against speculation and land grabbing. We continue to fight for land and territories.

SEEDS, THE COMMONS AND WATER

Seeds are at the heart of food sovereignty. Hundreds of organizations worldwide are joining with us to implement the principle of the Peoples' Heritage Seeds Serving Humanity. Our challenge is now to continue keeping our living seeds in the hands of our communities, by multiplying seeds on our farms and territories. We continue to fight against the misappropriation of seeds through various forms of intellectual property and the contamination of stocks with GM technology. We oppose the distribution of technological packages combining GMO seeds with the massive use of pesticides.

We will continue to share seeds knowing that our knowledge, our science, our practice as guardians of seed diversity are crucial to adapting to climate change.

The cycles of life flow through water. Water is an essential part of ecosystems and all life. Water is a commons and therefore it must be protected.

Building on our strengths

Our strength is creating and maintaining unity through diversity. We present our vision which is inclusive, broadly based, practical, radical and hopeful as an invitation to join us in transforming our societies and protecting Mother Earth.

- Popular mobilization, confrontation with the powerful, active resistance, internationalism and local grassroots engagement are necessary components for effecting social change.
- In our courageous struggles for food sovereignty we continue to build essential strategic alliances with social movements, including workers, urban organizations, immigrants, groups resisting mega-dams and the mining industry, among others.
- Our main tools are training, education and communication. We are exchanging our accumulated knowledge of the methods and content of cultural, political, ideological and technical training. We are multiplying our schools, educational experiences and communication instruments with our bases.
- We are committed to creating empowering spaces for rural youth. Our greatest hope for the future is the passion, energy and commitment articulated in the youth in our movement.

We go forward from this VI International Conference of La Vía Campesina, embracing new organizations, confident in our strengths and filled with hope for the future.

Bibliography

Adams, B. and Tobin, K. 2015. *Confronting Development: A Critical Assessment of the UN's Sustainable Development Goals*, New York: Rosa Luxemburg Stiftung.

Albrow, M. 1996. *The Global Age*. Cambridge: Global Polity Press.

Alkire, S. 2005. Capability and Functionings: Definition and Justification. *Human Development and Capability Association*. 1 September 2005.

Alvarez, S.E., Dagnino, E. and Escobar, A. (1998). *Cultures of Politics/Politics of Cultures: Re-Visioning Latin American Social Movements*. Boulder: Westview Press.

Anthes, C. (forth.) *Contested Land – Contested Rights: The Transformation of Land Issues into a Matter of Human Rights at UN Level in Rome and Geneva*, PhD dissertation. Frankfurt: Peace Research Institute.

Araghi, F. 1995. Global Depeasantization: 1945–1990. *Sociological Quarterly*, 36(2): 337–368.

Ashcroft, B., Griffiths, G. and Tiffin, H. 1998. *Post-colonial Studies: Key Concepts*, London: Routledge, 159–160.

Avruch, K. and Black, P.W. 1993. Conflict Resolution in Intercultural Settings in D.J.D. Sandole and H. van der Merwe (eds), *Conflict Resolution Theory and Practice*, Manchester: Manchester University Press, 131–145.

Banks, N., Hulme, D. and Edwards, M. 2015. NGOs, States and Donors Revisited: Still too Close for Comfort?, *World Development*, 66: 707–718.

Beck, U. 2000. *What Is Globalization?* (Translated by P. Camiller) Cambridge: Polity Press.

Beck, U. and Cronin, C. 2006. *The Cosmopolitan Vision*. Cambridge: Polity Press.

Bellier, I. 2013. We Indigenous Peoples. Global Activism and the Emergence of a New Collective Subject at the United Nations, in B. Müller (ed.), *The Gloss of Harmony. The Politics of Policy-making in Multilateral Organisations*. London: Pluto Press, 177–201.

Bennet, L. 2003. Communicating Global Activism: Strengths and Vulnerabilities of Networked Politics, *Information, Communication & Society* 6(2): 143–168.

Berry, F. and Berry, W. 1999. Innovation and Diffusion Models in Policy Research, in P. Sabatier (ed.), *Theories of the Policy Process*, Boulder, CO: Westview.

Beuchelt, T.D. and Virchow, D. 2012. Food Sovereignty or the Human Right to Adequate Food: Which Concept Serves Better as International Development Policy for Global Hunger and Poverty Reduction? *Agriculture and Human Values*, 29(2): 259–273.

Blumer, H. 1969. Collective Behavior, in A.M. Lee (ed.), *Principles of Sociology*, 3rd edn, New York: Barnes and Noble Books.

Bohm, S., Dinerstein, A.C. and Spicer, A. 2010. (Im)possibilities of Autonomy. Social Movements in and Beyond Capital, the State and Development, *Social Movement Studies*, 9 (1): 17–32.

Bohman, J. 1999. Democracy as Inquiry, Inquiry as Democratic: Pragmatism, Social Science, and the Cognitive Division of Labor, *American Journal of Political Science*, 43(2): 590–607.

Borras, S. 2004. *La Vía Campesina. An Evolving Transnational Social Movement*, TNI Briefing Series No 6. Transnational Institute.

Borras, S. 2008. La Vía Campesina and its Global Campaign for Agrarian Reform, in S. Borras et al. (eds) *Transnational Agrarian Movements Confronting Globalization*, West Sussex, UK: Wiley-Blackwell, 91–122.

Borras, S. 2008b. *Transnational Agrarian Movements Confronting Globalization*. West Sussex, UK: Wiley-Blackwell.

Borras, S. 2010. From Threat to Opportunity? Problems with the Idea of a 'Code of Conduct' for Land-Grabbing, *Yale Human Rights & Development Law Journal* 13: 507–523.

Borras, J. 2016. Land Politics, Agrarian Movements and Scholar-activism, Inaugural Lecture, 14 April 2016: International Institute of Social Studies: Available: https://www.tni.org/files/publication-downloads/borras_inaugural_lecture_14_april_2016_final_formatted_pdf_for_printing.pdf

Borras, S., Edelman, M. and Kay, C. 2008. Transnational Agrarian Movements: Origins and Politics, Campaigns and Impact, in S. Borras et al. (eds) *Transnational Agrarian Movements Confronting Globalization*, West Sussex, UK: Wiley-Blackwell, 1–36.

Borras, S. and Franco, J.C. 2009. *Transnational Agrarian Movements. Struggling for Land and Citizenship Rights*, IDS Working Paper 323. Institute of Development Studies, Brighton, England, April.

Boudon, R. and Bourricaud, F. 1986. *Dictionnaire critique de la sociologie*, Paris: Presses Universitaires de France.

Bové, J. 2001. A Farmers' International. *New Left Review* 12, November/December 2001.

Bové, J. and Dufour, F. 2001. *The World is Not For Sale: Farmers Against Junk Food*, London: Verso.

Brehm, V.M. et al. 2004. *Autonomy or Dependence? Case Studies of North–South NGO Partnerships*. Oxford: INTRAC.

Brem-Wilson, J. 2010. The Reformed CFS – A Briefing Paper for Civil Society. Facilitated by the IPC – International Planning Committee for Food Sovereignty, November, 2010.

Brem-Wilson, J. 2011. *La Via Campesina and the Committee on World Food Security: A Transnational Public Sphere?* (PhD Thesis) International Centre for Participation Studies, Department of Peace Studies, University of Bradford.

Brem-Wilson, J. 2014. From 'Here' to 'There': Social Movements, the Academy and Solidarity Research, *Socialist Studies/Études socialistes*, 10(1): 111–132.

Brem-Wilson, J. 2015. Towards Food Sovereignty: Interrogating Peasant Voice in the UN Committee on World Food Security, *Journal of Peasant Studies*, 42(1): 73–95.

Bringel, B. and Domingues, J.M. 2012. Teoría crítica e movimentos sociais: Intersecções, impasses e alternativas, in B. Bringel and M. da Glória Gohn (eds), *Movimentos sociais na era global*, Rio de Janeiro/Petrópolis: Vozes, pp. 57–76.

Bringel, B. 2015. Social Movements and Contemporary Modernity: Internationalism and Patterns of Global Contestation, in B. Bringel and J.M. Domingues (eds), *Global Modernity and Social Contestation*, London/New Delhi: Sage.

Bringel, B. and Domingues, J.M. (eds) 2015. Introduction, *Global Modernity and Social Contestation*, London/New Delhi: Sage.

Brinkerhoff, D.W. and Goldsmith, A. 2002. *Clientelism, Patrimonialism and Democratic Governance: An Overview and Framework for Assessment and Programming*, Washington.

Report prepared for US Agency for International Development, Office of Democracy and Governance.
Buchanan, A. and Keohane, R.O. 2006. The Legitimacy of Global Governance Institutions, *Ethics & International Affairs*, 20(4): 405–437.
Burawoy, M. (ed.). 2000. *Global Ethnography: Forces, Connections, and Imaginations in a Post-modern World*, Berkeley: University of California Press.
Burt, R.S. 2005. *Brokerage and Closure: An Introduction to Social Capital*, Oxford: Oxford University Press.
Böhm, S., Dinerstein, A.C. and Spicer, A. 2010. (Im)possibilities of Autonomy: Social Movements in and beyond Capital, the State and Development, *Social Movement Studies*, 9 (1): 17–32
Cardoso, F.H. 2003. Civil Society and Global Governance: Contextual Paper: http://www.unece.org/fileadmin/DAM/env/pp/ppif/Civil%20Society%20and%20Global%20Governance %20paper%20by%20Cardoso.htm
Caruso, G. 2013. Transformative Ethnography of the World Social Forum: Theories and Practices of Global Transformations, in Juris, J.S. and Khasnabish, A. (eds), *Insurgent Encounters: Transnational, Activism, Ethnography, and the Political*, Durham, NC: Duke University Press.
Caruso, G. 2014. Beyond the Square: Changing Dynamics at the World Social Forum, *Open Democracy*, 7 December 2014.
Castells, M. 1996. *The Information Age. Economy, Society and Culture*, Oxford, Malden, MA: Blackwell.
Castells, M. 2000. *End of Millennium: The Information Age: Economy, Society and Culture*, Vol. 3, 2nd edition, Oxford: Blackwell.
CFS2009. Reform of The Committee on World Food Security October 2009: http://www.fao.org/fileadmin/templates/cfs/Docs0910/ReformDoc/CFS_2009_2_Rev_2_E_K7197.pdf
CFS2010. Annex 3: Key Principles Coming out of the People's Food Sovereignty Forum: Civil Society's Role in Global Food Governance: http://www.cigionline.org/articles/2009/12/people's-food-sovereignty-forum-civil-society's-role-global-food-governance
CFS2011. Committee on World Food Security Brochure: http://www.fao.org/fileadmin/templates/cfs/Docs1112/Brochure/CFS_Brochure_2011_En.pdf
Chambers, R. 1994. Paradigm Shifts and the Practice of Participatory Research and Development. IDS Working Paper no. 2. Brighton: IDS.
Christine, O. 2012. The Effect of Global Economic Crisis on Service Delivery in Selected Non-Governmental Organizations in Kenya, *Journal of Management and Strategy* 2(4) December.
Civil Society Mechanism (CSM). 2010a. Proposal for an International Food Security and Nutrition Civil Society Mechanism for Relations with the CFS.
Civil Society Mechanism (CSM). 2010b. DRAFT Guidelines for facilitating common policy positions and messages through the Civil Society Mechanism: http://www.csm4cfs.org/wp-content/uploads/2016/03/draft_guidelines_on_common_policy_positions_en.pdf (Accessed 1 October 2016).
Civil Society Mechanism (CSM). 2012. CSM Annual Report 2012: http://www.csm4cfs.org/wp-content/uploads/2016/03/csm_annual_report_2012.pdf (Accessed 1 October 2016).
Civil Society Mechanism (CSM). 2014a. Evaluation. Conducted by Christina Schiavoni and Patrick Mulvany. Final Report, August 2014.

Civil Society Mechanism (CSM). 2014b. Actions and Policies on Investment in Agriculture: www.csm4cfs.org/files/SottoPagine/118/actions_and_policies_proposals_from_csm_en.pdf

Civil Society Mechanism (CSM). 2015. CSM Working Groups (homepage) http://www.csm4cfs.org/policy-working-groups/ (Accessed 1 October 2016).

CIVICUS. 2014. Open Letter to our Fellow Activists across the Globe: Building from Below and Beyond Borders, 6 August 2014: http://blogs.civicus.org/civicus/2014/08/06/an-open-letter-to-our-fellow-activists-across-the-globe-building-from-below-and-beyond-borders/#more-1750 (Accessed 1 October 2016).

Claeys, P. 2012. The Creation of New Rights by the Food Sovereignty Movement: The Challenge of Institutionalizing Subversion, *Sociology* 46(5): 844–860.

Claeys, P. 2013. Claiming Rights and Reclaiming Control: The Creation of New Human Rights by the Transnational Agrarian Movement Vía Campesina and the Transformation of the Right to Food (PhD dissertation). University of Louvain.

Claeys, P. 2015. Food Sovereignty and the Recognition of New Rights for Peasants at the UN: A Critical Overview of La Via Campesina's Rights Claims over the Last 20 Years, *Globalizations*, 12(4): 452–465.

Clapp, J. and D. Fuchs. 2009. Agri-food Corporations, Global Governance, and Sustainability: A Framework for Analysis, in J. Clapp and D. Fuchs (eds), *Corporate Power in Agri-food Governance*, London: The MIT Press, 1–26.

Cornwall, A. and Coelho, V.S. 2007. Spaces for Change? The Politics of Participation in New Democratic Arenas, in L. Thompson and C. Tapscott (eds), *Citizenship and Social Movements: Perspectives from the Global South*, Chicago: Chicago University Press.

Cortez, D. 2011. *La Constrcucción Social del Buen Vivir en Ecuador*, Quito: PADH.

Cotula, L. 2011. Land Deal in Africa: What is in the Contracts? International Institute for Environment and Development (UK), January: http://pubs.iied.org/pdfs/12568IIED.pdf (Accessed 1 October 2016).

Cox, L. and Fominaya, C.F. 2009. Movement Knowledge: What Do We Know, How Do We Create Knowledge and What Do We Do With It? *Interface: A Journal for and about Social Movements*, 1(1): 1–20.

Croteau, D. 2005. Which Side Are You On? The Tension Between Movement Scholarship and Activism, in Croteau, D., Haynes, W. and Ryan, C. (eds), *Rhyming Hope and History: Activists, Academics and Social Movement Scholarship*, Minneapolis: University of Minneapolis Press.

Dagnino, E. 2002. *Sociedade Civil e Espacos Publicos no Brasil*, Sao Paulo: Paz e Terra.

Dagnino, E. 2010. Civil Society in Latin America: Participatory Citizens or Service Providers? In H. Moksnes and M. Melin (eds), *(Con-)Tested Civil Society in Search of Democracy*, Uppsala: Uppsala Universitet.

Dagnino, E. and TatagibaL. 2010. Mouvements sociaux et participation institutionnelle: répertoires d'action collective et dynamiques culturelles dans la difficile construction de la démocratie brésilienne. *Revue internationale de politique comparée*, 2010/2, 17: 167–185.

Dawn, S.S.K. 2008. *Consciousness or Co-optation: Ethnic Political Power and Movement Outcomes in Ecuador and Australia*, Northwestern University.

De Certeau, M. 1990. *L'invention du quotidien. L'Arts de faire*, Paris: Gallimard.

De Certeau, M. 1984. *The Practice of Everyday Life*, Berkeley: University of California Press.

De Schutter, O. 2009. A New Role for the Committee on World Food Security. Reform note, meeting of the Contact Group to support the Committee on World

Food Security (CFS) 22 May 2009: www2.ohchr.org/english/issues/food/docs/CFS_reform_note22May09.pdf2nd

De Schutter, O. 2010. Responsibly Destroying the World's Peasantry: Land Grabbing's Grim Reality, in *Right to Food Watch, Land Grabbing and Nutrition: Challenges for Global Governance*: www.rtfn-watch.org/fileadmin/media/rtfn-watch.org/ENGLISH/pdf/Watch_2010/watch_engl_innen_final_a4.pdf

De Schutter, O. 2011. How Not to Think of Land-grabbing: Three Critiques of Large-scale Investments in Farmland, *Journal of Peasant Studies*, 38(2): 249–279.

De Schutter, O. 2013a. The FAO Must Do More to Promote Food as a Basic Human Right. The UN's Food and Agriculture Organisation needs to accelerate its movement towards a rights-based approach to food security, *The Guardian*, 4 March 2013: https://www.theguardian.com/global-development/poverty-matters/2013/mar/04/fao-food-basic-human-right (Accessed 1 October 2016).

De Schutter, O. 2013b. Report of the Special Rapporteur on the Right to Food, Olivier De Schutter Addendum Mission to the Food and Agriculture Organization of the United Nations: http://www.srfood.org/images/stories/pdf/officialreports/20130304_fao_en.pdf (Accessed 1 October 2016).

Della Porta, D. 2008. *Consensus in Movements*, Florence: European University Institute.

Della Porta, D., Andretta, M., Mosca, L. and Reiter, H. 2006. *Globalization from Below: Transnational Activists and Protest Networks*, Minneapolis: University of Minnesota Press.

Della Porta, D. and Diani, M. 1999. *Social Movements: An Introduction*, Oxford: Blackwell.

Desmarais, A.A. 2002. The Vía Campesina: Consolidating an International Peasant and Farm Movement, *Journal of Peasant Studies* 29(2): 91–124.

Desmarais, A.A. 2007. *La Vía Campesina: Globalization and the Power of Peasants*, Halifax and London: Fernwood Publishing and Pluto Press.

Desmarais, A.A. 2008. The Power of Peasants: Reflections on the Meanings of La Via Campesina, *Journal of Rural Studies* 24(2): 138–149.

Desmarais, A.A. and Nicholson, P. 2013. La Via Campesina: An Historical and Political Analysis, *La Via Campesina's Open Book: Celebrating 20 Years of Struggle and Hope*: http://www.viacampesina.org/en/index.php/publications-mainmenu-30/1409-la-via-campesina-s-open-book-celebrating-20-years-of-struggle-and-hope (Accessed 1 October 2016).

Doerr, N. 2012. Translating Democracy: How Activists in the European Social Forum Practice Multilingual Deliberation, *European Political Science Review* 4(03): 361–384.

Doucet, M.G. 2008. World Politics, the Alter-globalization Movement and the Question of Democracy, in M. Hammond-Calaghan and M. Hayday (eds), *Mobilizations, Protests and Engagements: Canadian Perspectives on Social Movements*, Black Point (Nova Scotia): Fernwood Publishing, 18–34.

Dryzek, J.S., Bächtiger, A. and Milewicz, K. 2011. Towards a Deliberative Global Citizens' Assembly, *Global Policy* 2(1): 33–43.

Duncan, J. 2014. The Reformed Committee on World Food Security and the Global Governance of Food Security. PhD dissertation.

Duncan, J. 2015. *Global Food Security Governance: Civil Society Engagement in the Reformed Committee on World Food Security*, London: Routledge.

Duncan, J. and Barling, D. 2012. Renewal through Participation in Global Food Security Governance: Implementing the International Food Security and Nutrition Civil Society Mechanism to the Committee on World Food Security, *International Journal of Sociology of Agriculture and Food* 19(2).

Dwyer, S.C. and Buckle, J.L. 2009. The Space Between: On Being an Insider-Outsider in Qualitative Research, *International Journal of Qualitative Methods* 8(1): 55–63.

Edelman, M. and Carwil, J. 2011. Peasants' Rights and the UN System: Quixotic Struggle? Or Emancipatory Idea whose Time has Come? *Journal of Peasant Studies* 38(1): 81–108.

Edelman, M. 1999. *Peasants against Globalization: Rural Social Movements in Costa Rica*, Stanford: Stanford University Press.

Edelman, M. 2003. Transnational Peasant and Farmer Movements and Networks, in M. Kaldor, H. Anheier and M. Glasius (eds), *Global Civil Society*, London: Oxford University Press, pp. 185–220.

Edelman, M. 2008. Transnational Organizing in Agrarian Central America: Histories, Challenges, Prospects, *Journal of Agrarian Change* 8(2/3): 229–257.

Edelman, M. 2009. Synergies and Tensions between Rural Social Movements and Professional Researchers, *Journal of Peasant Studies*, 36(1): 245–265.

Edelman, M. 2012. One-third of Humanity: Peasant Rights in the United Nations, *OpenDemocracy*, 10 October: https://www.opendemocracy.net/marc-edelman/one-third-of-humanity-peasant-rights-in-united-nations (Accessed 1 October 2016).

Edwards, M. and Gaventa, J. 2001. *Citizen Global Action*, Boulder: Lynne Rienner Press.

Epstein, E. 1991. *Political Protest and Cultural Revolution*, Berkeley: University of California Press.

Eschele, C. 2001. Globalizing Civil Society. Social Movements and the Challenge of Global Politics from Below, in P. Hamel, H. Lustiger-Thaler, J.N. Pieterse and S. Roseneil (eds), *Globalization and Social Movements*, Houndmills, Basingstoke, Hampshire: Palgrave.

Escobar, A. 1995. *Encountering Development: The Making and Unmaking of the Third World*, Princeton, New Jersey: Princeton University Press.

Escobar, A. 2001. Culture Sits in Places: Reflections on Globalism and Subaltern Strategies of Globalization, *Political Geography*, 20: 139–174.

Escobar, A. 2004. Beyond the Third World: Imperial Globality, Global Coloniality and Anti-Globalisation Social Movements, *Third World Quarterly* 25(1): 207–230.

Escobar, A. 2008. *Territories of Difference: Place, Movements, Life, Redes*, London: Duke University Press.

Everett, K. 1992. Professionalization and Protest: Changes in the Social Movement Sector, 1961–1983, *Social Forces* 70(4): 957–975.

Fairclough, N. 1992. *Discourse and Social Change*. London: Polity Press, Palgrave.

FAO. 2008. Committee on World Food Security: Participation of Civil Society/Non-Governmental Organizations. Document reference: CFS: 2008/6. ftp://ftp.fao.org/docrep/fao/meeting/014/k3028e.pdf (Accessed 1 October 2016).

FAO. 2011. Land Tenure and International Investments in Agriculture. A report by The High Level Panel of Experts on Food Security and Nutrition, July 2011.

FAO. 2012a. Respecting Free, Prior and Informed Consent. Practical Guidance for Governments, Companies, NGOs, Indigenous Peoples and Local Communities in Relation to Land Acquisition: http://www.fao.org/3/a-i3496e.pdf (Accessed 1 October 2016).

FAO. 2012b. Countries Adopt Global Guidelines on Tenure of Land, Forests, Fisheries. 11 March 2012. http://www.fao.org/news/story/en/item/142587/icode/ (Accessed 1 October 2016).

FAO. 2013. FAO Recognizes La Via Campesina's Crucial Role as the Major International Small Food Producer's Organization. 4 October 2014: http://viacampesina.

org/en/index.php/main-issues-mainmenu-27/food-sovereignty-and-trade-mainmenu-38/1497-fao-recognizes-la-via-campesina-s-crucial-role-as-the-major-international-small-food-producer-s-organisation (Accessed 1 October 2016).

FAO. 2014a. Respecting Free, Prior and Informed Consent: Practical Guidance for Governments, Companies, NGOs, Indigenous Peoples and Local Communities in Relation to Land Acquisition; http://www.fao.org/3/a-i3496e.pdf (Accessed 1 October 2016).

FAO. 2014b. The Principles for Responsible Investment in Agriculture and Food Systems, 41st Session of CFS, 15 October: http://www.fao.org/3/a-ml291e.pdf (Accessed 1 October 2016).

FAO et al. 2010. The Seven Principles for Responsible Agricultural Investment that Respects Rights, Livelihoods and Resources. A discussion note prepared by FAO, IFAD, UNCTAD and the World Bank Group to contribute to an ongoing global dialogue.

Ferguson, J. 1994. *The Anti-Politics Machine: 'Development,' Depoliticization, and Bureaucratic Power in Lesotho*, Cambridge: Cambridge University Press.

FIAN 2010. Why We Oppose the Principles for Responsible Agricultural Investment (Rai), The Global Campaign for Agrarian Reform Land Research Action Network, October: www.fian.org/fileadmin/media/ publications/2010_09_Oppose_RAI.pdf (Accessed 1 October 2016).

FIAN et al. 2011. Dakar Appeal against the Land Grab: http://fian.org/news/press-releases/g20-agriculture-hundreds-of-organizations-say-stop-farm-land-grabbing (Accessed 1 October 2016).

Flacks, R. 2004. Knowledge for What? Thoughts on the State of Social Movement Studies, in Goodwin, J. and Jasper, J. (eds), *Rethinking Social Movements: Structure, Culture, and Emotion*, Lanham, MD: Rowman and Littlefield.

Fox, J. 1993. *The Politics of Food in Mexico: State Power and Social Mobilization*, Ithaca, NY: Cornell University Press.

Fox, J. 1994. The Difficult Transition from Clientelism to Citizenship: Lessons from Mexico, *World Politics* 46(2): 151–184.

Fox, J. 2000. The World Bank Inspection Panel: Lessons from the First Five Years, *Global Governance*, 6: 279–318.

Fox, J. 2003. Framing the Inspection Panel, in D. Clark, J. Fox and K. Treacle (eds) *Demanding Accountability: Civil-Society Claims and the World Bank Inspection Panel*, Lanham, MD: Rowman and Littlefield, pp. xi–xxxi.

Freeman, J. 1972. Tyranny of Structurelessness, *Berkeley Journal of Sociology* 17, 1972–1973: 151–165.

Freire, P. 1970. *Pedagogy of the Oppressed*, New York: Continuum.

Freire, P. 1973. *Education as the Practice of Freedom in Education for Critical Consciousness*, New York: Continuum.

Friedman, J. 1999. Indigenous Struggles and the Discreet Charm of the Bourgeoisie, *The Australian Journal of Anthropology* 10(1): 1–14.

Fung, A. and Wright, E.O. 2001. Deepening Democracy: Innovations in Empowered Participatory Governance, *Politics & Society*, 29(1): 5–41.

Gale, R.P. 1986. Social Movements and the State: The Environmental Movement, Counter-movement, and Governmental Agencies, *Sociological Perspectives* 29: 202–240.

Gamson, W.A. 1990. *The Strategy of Social Protest*, 2nd edition. Belmont, CA: Wadsworth Publishing.

Gaarde, I. and Hoegh-Jeppesen, M. 2011. The Trend of Land Sales and the Challenges in order to Meet the United Nations' Right to Food. Master's thesis. University of Copenhagen.

Gaunlett, D. 2002. *Media Gender and Identity: About Giddens' Work on Modernity and Self-identity*, Routledge: London.

Gaventa, J. 2006. Finding the Spaces for Change: A Power Analysis, *IDS Bulletin* 37(6): 23–33.

George, R. 1993. *The McDonaldization of Society*, Pine Forge Press.

Giddens, A. 1984. *The Constitution of Society: Outline of the Theory of Structuration*, University of California Press.

Giddens, A., and Pierson, C. 1998. *Conversations with Anthony Giddens: Making sense of modernity*. Stanford, CA: Stanford University Press.

Gillan, K., Pickerill, J. et al. 2012. Special Issue: *The Ethics of Research on Activism. Social Movement Studies: Journal of Social, Cultural and Political Protest*, 11(2).

Gitz, V. and Meybeck, A. 2011. The Establishment of the High Level Panel of Experts on Food Security and Nutrition, CIRED Working Papers, 2011-30.

Giugni, M. and Passy, F. 1998. Social Movements and Policy Change: Direct, Mediated, or Joint Effect?, American Sociological Association's Section on Collective Behavior and Social Movements Working Paper Series, Vol. 1.

GRAIN. 2011. Não há justificativas para a grilagem legalizada de terras, 5 October 2011.

Gramsci, A. 1971. *Selections from the Prison Notebooks*. Trans. and ed. Q. Hoare and G.N. Smith, New York: International Publishers.

Guiso, A. 2000. Potenciando la Diversidad. Dialogo de saberes, una practica hermeneutica colectiva, *Revista Aportes* (53): 57–70.

Gwynne, N. and Kay, C. 2004 (eds) *Latin America Transformed: Globalization and Modernity*, 2nd edition, London: Arnold,

Habermas, J. 1997. *Droit et démocratie. Entre faits et normes*, Paris: Gallimard.

Hardt, M. and Negri, A. 2004. *Multitude: War and Democracy in the Age of Empire*, Penguin, 2004.

Heller, P. 2013. Civil Society and Social Movements in a Globalizing World, Human Development Report Office, Occasional Papers 2013/06.

Hitchman, J. 2015. Who Controls the Food System? Essay submitted as part of TNI's call for papers for its *State of Power* 2015 report.

Holt-Giménez, E. and Patel, R. 2009. *Food Rebellions: Crisis and the Hunger for Justice*, Pambazuka Press.

Hyman, R. 2010. *The International Labour Movement on the Threshold of Two Centuries of Agitation, Organisation, Bureaucracy, Diplomacy*, London.

Inglehart, R. 1997. *Modernization and Postmodernization: Cultural, Economic, and Political Change in 43 Societies*, Princeton, NJ: Princeton University Press.

James, J. and Goodwin, J. 1999. Caught in a Winding, Snarling Vine: the Structural Bias of Political Process Theory, *Sociological Forum* 14(1): 27–54.

Jasper, J. 2004. A Strategic Approach to Collective Action: Looking for Agency in Social-movement Choices, *Mobilization* 9: 1–16.

Jasper, J. 2010a. Social Movement Theory Today: Toward a Theory of Action? *Sociology Compass* 10.

Jasper, J. 2010b. Strategic Marginalizations and Emotional Marginalities: The Dilemma of Stigmatized Identities, in D.S. Roy *Surviving Against Odds: The Marginalized in a Globalizing World*, New Delhi: Manohar, 29–37.

Jasper, J. 2014. Playing the Game: Introduction to Players and Arenas. In J.M. Jasper and J.W. Duyvendak, *Players and Arenas. The Interactive Dynamics of Protests*, Amsterdam University Press.

Jasper, J. and Polletta, F. 2001. Collective Identity and Social Movements, *Annual Review of Sociology* 27: 283–305.

Juris, J.S. 2007. Practicing Militant Ethnography with the Movement for Global Resistance in Barcelona, in Shukaitis, S. and Graeber, D. (eds), *Constituent Imagination: Militant Investigations, Collective Theorization*, Oakland, CA: AK Press, pp. 164–176.

Juris, J.S. and Khasnabish, A. (eds) 2013. *Insurgent Encounters: Transnational, Activism, Ethnography, and the Political*, Durham, NC: Duke University Press.

Kaldor, M. 2003. *Global Civil Society: An Answer to War*, Cambridge: Polity.

Kaldor, M. 2012. Global Civil Society 2012: Ten Years of 'Politics from Below', *OpenDemocracy*, 30 April.

Karatzogianni, A. 2006. *The Politics of Cyberconflict*, London and New York, NY: Routledge.

Karns, M. and Mingst, K. (eds). 2010. *International Organizations: The Politics and Processes of Global Governance*, Boulder: Lynne Rienner Publishers.

Kaufmann, J.C. 1996. *L'entretien compréhensif*, Collection 128. Sociologie, Paris: Nathan.

Keck, M. and Sikkink, K. 1998. *Activists Beyond Borders: Advocacy Networks in Transnational Politics*, Ithaca: Cornell University Press.

Kinder, D.R. 1998. Opinion and Action in the Realm of Politics, in D.T. Gilbert, S.T. Fiske, and G. Lindzey (eds) *The Handbook of Social Psychology*, 4th edition, Vol. 2, New York: Oxford University Press, 778–867.

Kitschelt, H. 1986. Political Opportunity Structures and Political Protest: Anti-Nuclear Movements in Four Democracies, *British Journal of Political Science* 16: 57–85.

Kriesi, H. 1996. The Organizational Structure of New Social Movements in a Political Context, in D. McAdam, J.D. McCarthy, M.N. Zald *Comparative Perspectives on Social Movements: Political Opportunities, Mobilizing Structures, and Cultural Framings*, Cambridge: Cambridge University Press.

Kurzman, C. 2011. Meaning Making in Social Movements, *Anthropology Quarterly*, 81(1): 17–58.

La Vía Campesina (LVC). 1993. Declaration Mons. May.

La Vía Campesina (LVC). 1996. The Right to Produce and Access to Land. Food Sovereignty: A Future without Hunger. 11–17 November 1996 in Rome, Italy: http://www.acordinternational.org/silo/files/decfoodsov1996.pdf Accessed 10 March 2016.

La Vía Campesina (LVC). 2004a. Policy Documents. Maputo. 2008. http://viacampesina.org/downloads/pdf/en/EN-policy-document.pdf

La Vía Campesina (LVC). 2004b. Radical Opposition to Neoliberalism. *Bulletin* 5. http://viacampesina.org/en/index.php/our-conferences-mainmenu-28/4-sao-paolo-2004-mainmenu-43/338-radical-opposition-to-neoliberalism (Accessed 19 March 2015).

La Vía Campesina (LVC). 2006. Seattle Declaration: Take WTO out of Agriculture: https://viacampesina.org/en/index.php/actions-and-events-mainmenu-26/10-years-of-wto-is-enough-mainmenu-35/43-seattle-declaration-take-wto-out-of-agriculture (Accessed 1 October 2016).

La Vía Campesina (LVC). 2008. Background and policy documents. 5th International Conference in Maputo, Mozambique: https://viacampesina.org/downloads/pdf/en/EN-policy-document.pdf (Accessed 1 October 2016).

La Vía Campesina (LVC). 2009. Declaration of Rights of Peasants – Women and Men. https://viacampesina.net/downloads/PDF/EN-3.pdf (Accessed 1 October 2016).

La Vía Campesina (LVC). 2010. Nourishing the Planet. Fighting for Food Sovereignty, Social Justice, Land Rights and Gender Equity. 19 August. (Accessed 14 August 2014).

La Vía Campesina (LVC). 2012a. The Reform of The Committee on World Food Security: A New Space for Policies of the World, Opportunities and Limitations. 2 November. https://viacampesina.org/downloads/pdf/en/report-no.4-EN-2012-comp.pdf (Accessed 14 September 2014).

La Vía Campesina (LVC). 2012b. La Via Campesina in the Committee on World Food Security: Investments Needed for Small Scale Farming, Not for Agribusiness. 11 October 2012. https://viacampesina.org/en/index.php/main-issues-mainmenu-27/food-sovereignty-and-trade-mainmenu-38/1310-la-via-campesina-in-the-committee-on-world-foodsecurity-investments-needed-for-small-scale-farming-not-foragribusiness (Accessed 24 October 2013).

La Vía Campesina (LVC). 2012c. Peasants of the World Mobilize against Green Capitalism in Rio. 15 June 2012. https://viacampesina.org/en/index.php/actions-and-events-mainmenu-26/-climate-change-and-agrofuels-mainmenu-75/1248-the-people-of-the-world-confront-the-advance-of-capitalism-rio-20-and-beyond (Accessed 1 September 2014).

La Vía Campesina (LVC). 2012d. Why Are the FAO and the EBRD Promoting the Destruction of Peasant and Family Farming? 12 September 2012. https://viacampesina.org/en/index.php/main-issues-mainmenu-27/agrarian-reform-mainmenu-36/1295-why-are-the-fao-and-the-ebrd-promoting-the-destruction-of-peasant-and-family-farming (Accessed 1 October 2014).

La Vía Campesina (LVC). 2013a. Declaration of the Vía Campesina 6th International Conference: The Jakarta Call. 20 December 2013 https://viacampesina.org/en/index.php/our-conferences-mainmenu-28/6-jakarta-2013/resolutions-and-declarations/1428-the-jakarta-call (Accessed 30 June 2013).

La Vía Campesina (LVC). 2013b. We Are Not Backward. Our Seeds and Knowledge Are Crucial for Our Survival. Biodiversity and Genetic Resources. 13 November 2013. https://viacampesina.org/en/index.php/main-issues-mainmenu-27/biodiversity-and-genetic-resources-mainmenu-37/1516-we-are-not-backward-our-seeds-and-knowledge-are-crucial-for-our-survival (Accessed 3 May 2014).

La Vía Campesina (LVC). 2014a. Report of La Vía Campesina 6th International Conference. http://viacampesina.org/downloads/pdf/en/EN-VITHCONF-2014.pdf (Accessed 7 May 2014).

La Vía Campesina (LVC). 2014b. Speech of Elisabeth Mpofu, general coordinator of La Via Campesina at the Yale Conference on Food Sovereignty. 24 January 2014.https://viacampesina.org/en/index.php/main-issues-mainmenu-27/food-sovereignty-and-trade-mainmenu-38/1560-via-campesina-at-the-colloquium-food-sovereignty-a-critical-dialogue (Accessed 3 May 2014).

La Vía Campesina (LVC). 2014c. CFS in Rome: The Majority of Governments Remain Blind to the Challenges of Global Food Security. 17 October 2014. https://viacampesina.org/en/index.php/main-issues-mainmenu-27/food-sovereignty-and-trade-mainmenu-38/1684-cfs-in-rome-the-majority-of-governments-remain-blind-to-the-challenges-of-global-food-security (Accessed 20 October 2014).

La Vía Campesina (LVC). 2014d. UN-masking Climate Smart Agriculture. 23 July 2014. https://viacampesina.org/en/index.php/main-issues-mainmenu-27/sustainable-peasants-agriculture-mainmenu-42/1670-un-masking-climate-smart-agriculture (Accessed 2 May 2015).

La Vía Campesina (LVC). 2015. The Guidelines on the Responsible Governance of Tenure at a Crossroads. 11 December 2015. https://viacampesina.org/en/index.php/main-issues-mainmenu-27/agrarian-reform-mainmenu-36/1933-the-guidelines-on-the-responsible-governance-of-tenure-at-a-crossroads (Accessed 1 June 2016).

Leff, E. 2004. Racionalidad ambiental y diálogo de saberes. Significancia y sentido en la construcción de un futuro sustentable, *Polis. Revista Latinoamericana* (7): 1–29.

Mann, A. 2014. *Global Activism in Food Politics: Power Shift*, Houndsmill: Palgrave Macmillan.

Marx, C. 1847. *The Poverty of Philosophy*, Chicago: C.H. Kerr and Co.

McAdam, D., McCarthy, J.D. and Zald, M.N. 1996. *Comparative Perspectives on Social Movements: Political Opportunities, Mobilizing Structures, and Cultural Framings*, New York: Cambridge University Press.

Massicotte, M.J. 2010. La Via Campesina, Brazilian Peasants, and the Agribusiness Model of Agriculture: Towards an Alternative Model of Agrarian Democratic Governance, *Studies in Political Economy* 85: 69–98.

McCloskey, S. 2012. Aid, NGOs and the Development Sector: Is it Time for a New Direction?, *Policy & Practice: A Development Education Review* 15, Autumn.

McDonald, K. 2006. *Global Movements: Action and Culture*. Oxford: Blackwell.

McDonald, K. 2007. Between Autonomy and Vulnerability: The Space of Movement, *Recherches Sociologiques et Anthropologiques*, 38(1): 49–63.

McKeon, N. 2009a. *The United Nations and Civil Society: Legitimating Global Governance – Whose Voice?* London: Zed Books.

McKeon, N. 2009a. A Food Battle Won. http://www.csm4cfs.org/files/Pagine/16/a_food_battle_won_nora_mckeon_en.pdf (Accessed 1 October 2016).

McKeon, N. 2009b. Who Speaks for Peasants? Civil Society, Social Movements and the Global Governance of Food and Agriculture, *Interface: A Journal for and about Social Movements* 1(2): 48–82.

McKeon, N. 2010. Who Speaks for the Poor, and Why Does It Matter? *UN Chronicle* 47(3), November.

McKeon, N. 2013. One Does Not Sell the Land Upon Which the People Walk: Land Grabbing, Transnational Rural Social Movements, and Global Governance, *Globalizations*, 10(1): 105–122.

McKeon, N. 2015. *Food Security Governance: Empowering Communities, Regulating Corporations*, London: Routledge.

McKeon, N. and Carol, K. 2009. Strengthening Dialogue. UN Experience with Small Farmer Organizations and Indigenous Peoples, UN Non-Governmental Liaison Service (NGLS), October.

McMichael, P. 2005. Global Development and the Corporate Food Regime, in F. Buttel and P. McMichael (eds), *New Directions in the Sociology of Global Development*, 265–299. Amsterdam: Elsevier.

McMichael, P. 2006. Reframing Development: Global Peasant Movements and the New Agrarian Question, *Canadian Journal of Development Studies* 27(4): 471–483.

McMichael, P. 2008. Peasants Make Their Own History, But Not Just as They Please…, in S. Borras et al. (eds) *Transnational Agrarian Movements Confronting Globalization*, West Sussex, UK: Wiley-Blackwell.

McMichael, P. 2009. The World Food Crisis in Historical Perspective, *Monthly Review* 61(3) (August): 32–47.

McMichael, P. (ed.) 2010. *Contesting Development. Critical Struggles for Social Change*, New York and London: Routledge.

McMichael, P. 2015. The Land Question in the Food Sovereignty Project, *Globalizations: Food Sovereignty. Concept, Practice and Social Movements* 12(4): 434–451.

McMichael, P. and Schneider, M. 2011. Food Security Politics and the Millennium Development Goals, *Third World Quarterly*, 32(1), 119–139.

Melucci, A. 1996. *Challenging Codes: Collective Action in the Information Age*, Cambridge: Cambridge University Press.

Meyer, D.S. 1993. Political Process and Protest Movement Cycles: American Peace Movements in the Nuclear Age, *Political Research Quarterly* 46: 451–479.

Michels, R. [1915] 1962. *Political Parties: A Sociological Study of the Oligarchical Tendencies of Modern Democracy*, New York: Hearst's International Library.

Morgan, R. 2007. On Political Institutions and Social Movement Dynamics: The Case of the United Nations and the Global Indigenous Movement, *International Political Science Review* 28(3): 273–292.

Mouffe, C. 2000. *Deliberative Democracy or Agonistic Pluralism*, Political Science Series 72, December, Vienna: Institute for Advance Studies.

Movimento dos TrabalhadoresRuraisSem Terra. 2014. A agricultura camponesa e ecológica pode alimentar o mundo? 23 May: http://www.mst.org.br/node/16149 (Accessed 1 October 2016).

Müller, B. and Neveu, C. 2002. Mobilising Institutions – Institutionalizing Movements, *Focaal. European Journal of Anthropology*, special issue no. 40: 9–20.

Müller, B. 2011. The Elephant in the Room. Multi-stakeholder Dialogue on Agricultural Biotechnology in the FAO, in D. Però, C. Shore, S. Wright (eds) *Policy Worlds: Anthropology and the Anatomy of Contemporary Power*, Oxford: Berghahn Books.

Müller, B. (ed.) 2013. *The Gloss of Harmony. The Politics of Policy-making in Multilateral Organisations*, London: Pluto Press

Nyéléni newsletter. 2014. Creating Knowledge for Food Sovereignty. June 2014. http://www.nyeleni.org/spip.php?page=NWarticle.en&id_article=455 (Accessed 1 July 2014).

O'Brien, R., Goetz, A.M., Scholte, J.A. and Williams, M. 2000. *Contesting Global Governance: Multilateral Economic Institutions and Global Social Movements*. Cambridge: Cambridge University Press.

Offe, C. 1987. Challenging the Boundaries of Institutional Politics: Social Movements since the 1960s, in C. Maier (ed.) *Changing Boundaries of the Political*, Cambridge: Cambridge University Press.

Ortega-Espés, D., Highton, A.C., Strappazzon, Á., Pedot, E., Tzeiman, A., Icaza, M.S., and Seufert, P. 2016. *Manual Popular de las directrices voluntarias sobre la gobernanza responsable de la tenencia de la tierra, la pesca y los bosques en el contexto de la seguridad alimentaria nacional; Guía para la promoción, la aplicación, el monitoreo y la evaluación.* Comité Internacional de Planificación para la Soberanía Alimentaria.

Passy, F. 2003. *Social Movements Matter But How? Social Movements and Networks*, M. Diani and D. McAdam (eds), Oxford: Oxford University Press, pp.21–48.

Patel, R. 2006. International Agrarian Restructuring and the Practical Ethics of Peasant Movement Solidarity, *Journal of Asian and African Studies*, 41(1&2): 71–93.

Patel, R. 2009. What Does Food Sovereignty Look Like? *Journal of Peasant Studies*, 36(3): 663–706.

Peine, E. and McMichael, P. 2005. Globalization and Governance, in V. Higgins and G. Lawrence (eds), *Agricultural Regulation*, London: Routledge.

Piven, F. and Cloward, R. 1977. *Poor People's Movements: Why they Succeed, How they Fail*, New York: Vintage Books.

Pleyers, G. 2010. *Alter-globalization – Becoming Actors in the Global Age*, UK: Polity Press.
Pleyers, G. 2012. A Decade of World Social Forums: Internationalisation without Institutionalization, in M. Kaldor, H. Moore and S. Selchow (eds), *Global Civil Society*, London: Palgrave, pp. 166–182.
Pleyers, G. 2015. *The Global Age: A Social Movement Perspective in Global Modernity and Social Contestation*, London/New Delhi: Sage.
Pogge, T. 2002. *World Poverty and Human Rights*, Cambridge: Polity Press.
Poletta, F. 1998. Free Spaces in Collective Action, *Theory and Society*, 28: 1–38.
Poletta, F. 2002. *Freedom Is an Endless Meeting: Democracy in American Social Movements*, Chicago: University of Chicago Press.
Poletta, F. 2006. *It Was Like A Fever: Storytelling in Protest and Politics*, Chicago: University of Chicago Press.
Purwanto, H. 2013. Local To Global; How Serikat Petani Indonesia Has Accelerated The Movement For Agrarian Reform, in *La Via Campesinas Open Yearbook 2013*.
Randeria, S. 2007. The State of Globalization. Legal Plurality, Overlapping Sovereignties and Ambiguous Alliances between Civil Society and the Cunning State in India, *Theory, Culture &Society*, 24: 1–33.
Risse, T. 2002. Transnational Actors and World Politics, in W. Carlsnaes, T. Risse and B.A. Simmons (eds) *Handbook of International Relations*, London: Sage.
Rittberger, V. 2001. *Global Governance and the United Nations System*, New York: United Nations University Press.
Rosset, P. and Martínez, M.E. 2005. *Participatory Evaluation of La Vía Campesina*, The Norwegian Development Fund and La Via Campesina.
Rosset, P. and Martínez, M.E. 2010. La Vía Campesina: The Birth and Evolution of a Transnational Social Movement, *Journal of Peasant Studies*, 37(1).
Rosset, P. and Martínez, M.E. 2014. Diálogo de saberes in La Vía Campesina: Food Sovereignty and Agroecology, *The Journal of Peasant Studies*, 41(6): 979–997.
Routledge, P. 2004. A Relational Ethics of Struggle, in Fuller, D. and Kitchin, R. (eds), *Radical Theory, Critical Praxis: Making a Difference beyond the Academy?* Vernon and Victoria, BC: Praxis e-Press: 79–91.
Santos, A.C. 2012. Disclosed and Willing: Towards a Queer Public Sociology, *Social Movement Studies: Journal of Social, Cultural and Political Protest*, 11(2): 241–254.
Santos, B. de Sousa 2004. The World Social Forum: Toward a Counter-Hegemonic Globalisation, in J. Sen, A., Anand, A. Escobar and P. Waterman (eds), *World Social Forum: Challenging Empires*, New Delhi: Viveka.
Santos, B. de Sousa. 2007. Beyond Abyssal Thinking: From Global Lines to Ecologies of Knowledges, *Review*, XXX(1): 45–89.
Santos, B. de Sousa. 2012. Public Sphere and Epistemologies of the South, *African Development*, 37(1): 43–67.
Santos, B. de Sousa. 2014. *Epistemologies of the South: Justice Against Epistemicide*, Boulder: Paradigms.
Scholte, J.A. 2001. Civil Society and Democracy in Global Governance. CSGR Working Paper No. 65/01. Centre for the Study of Globalisation and Regionalisation.
Scholte, J.A. 2011. *Building Global Democracy? Civil Society and Accountable Global Governance*, Cambridge: Cambridge University Press.
Schumaker, P.D. 1975. Policy Responsiveness to Protest-Group Demands, *Journal of Politics*, 37. 488–521.
Scott, J. 1985. *Weapons of the Weak: Everyday Forms of Peasant Resistance*, New Haven and London: Yale University Press.

Sen, A. 1999. *Development as Freedom*, Oxford: Oxford University Press.

Schneiberg, M. and Lounsbury, M. 2008. Social Movements and Institutional Analysis, in R. Greenwood, C. Oliver, K. Sahlin-Andersson and R. Suddaby (eds), *The Handbook of Organizational Institutionalism*, London: Sage Publications, 650–672.

Smith, J. 2002. Bridging Global Divides? Strategic Framing and Solidarity in Transnational Social Movement Organizations, *International Sociology*, 17(4): 505–528.

Smith, J. 2008. *Social Movements for Global Democracy*, Baltimore, MD: John Hopkins University Press.

Smith, J. 2011. Transnational Activism and Global Social Change, in *Global Civil Society: Shifting Powers in a Shifting World*, Uppsala, December 2011.

Smith, J. and Wiest, D. 2012. *Social Movements in the World-System: The Politics of Crisis and Transformation*, New York: Russell Sage Foundation.

Snow, D.A. and Benford, R.D. 1988. Ideology, Frame Resonance and Participant Mobilization, in B. Klandermans, H. Kriesi and S. Tarrow (eds), *International Social Movement Research*, Vol. 1, Jai Press, 197–217.

Snow, D.A. and McAdam, D. 2000. Identity Work Processes in the Context of Social Movements: Clarifying the Identity/Movement Nexus, in S. Stryker, T.J. Owens and R.W. White (eds), *Self, Identity and Social Movements*, Minneapolis: University of Minnesota Press, 41–67.

Sorensen, E. and Torfing, J. 2005. The Democratic Anchorage of Governance Networks, *Scandinavian Political Studies*, 28(3): *195–218.*

Spivak, G. 1988. Subaltern Studies: Deconstructing Historiography, in R. Guha and G.C. Spivak (eds) *In Other Worlds: Essays in Cultural Politics*, Oxford: Oxford University Press, 197–221.

Staggenborg, S. 2002. Coalition Work in the Pro-Choice Movement: Organizational and Environmental Opportunities and Obstacles, *Social Problems*, 33(5): 374–390.

Steffek, J. and Ehling, U. 2008. Civil Society Participation at the Margins: The Case of the WTO, in J. Steffek, C. Kissling, and P. Nanz (eds), *Civil Society Participation in European and Global Governance. A Cure for the Democratic Deficit?* Basingstoke: Palgrave, 95–115.

Streck, C. 2002. Global Public Policy Networks as Coalitions for Change, in D.C. Esty and M.H. Ivanova (eds), *Global Environmental Governance, Options and Opportunitites*, New Haven: Yale School of Forestry and Environmental Studies, 121–140.

Tallberg, J. and Uhlin, A. 2011. Civil Society and Global Democracy: An Assessment, in D. Archibugi, M. Koenig-Archibugi and R. Marchetti (eds), *Global Democracy: Normative and Empirical Perspectives*, Cambridge: Cambridge University Press.

Tarrow, S. 1998. *Power in Movement: Social Movements and Contentious Politics*, Cambridge: Cambridge University Press.

Tarrow, S. 2001. Contentious Politics in a Composite Polity, in D. Imig and S. Tarrow (eds), *Contentious Europeans: Protest and Politics in an Emerging Polity*, Boulder, CO: Rowman & Littlefield, 233–251.

Tarrow, S. 2005. *The New Transnational Activism*, Cambridge: Cambridge University Press.

Taylor, V. and Van Dyke, N. 2004. 'Get up, Stand up': Tactical Repertoires of Social Movements, in D.A. Snow, S.A. Soule and H. Kriesi (eds), *The Blackwell Companion to Social Movements*, Sussex, UK: Wiley-Blackwell.

Tilly, C. 1978. *From Mobilization to Revolution*, Reading, MA: Addison-Wesley.

Tilly, C. 2004. *Social Movements, 1768–2008*, Boulder, CO: Paradigm Publishers.

Tilly, C. and Tarrow, S. 2007. *Contentious Politics*, Boulder, CO: Paradigm Publishers.

Touraine, A. 1973. *Production de la sociéte'*, Paris: Seuil.

Touraine, A. 1978. *La Voix et le regard*. Paris: Les Éditions du Seuil.

Touraine, A. 1981. *The Voice and the Eye: An Analysis of Social Movements*, Cambridge: Cambridge University Press.

Touraine, A. 1985. An Introduction to the Study of New Social Movements, *Social Research*, 52(4): 749–787.

Touraine, A. 2001. *Beyond Neoliberalism*, London: Polity Press.

Touraine, A. 2001. From Understanding Society to Discovering the Subject, *Anthropological Theory* 2(4): 387–398.

Touraine, A. 2007. *The Merging of Knowledge: People in Poverty and Academics Thinking Together*, Fourth World-University Research Group, University Press of America.

Touraine, A., Wieviorka, M. and Dubet, F. 1987. *The Workers' Movement*, Cambridge: Cambridge University Press.

Van der Ploeg, J.D. 2008. *The New Peasantries: Struggles for Autonomy and Sustainability in an Era of Empire and Globalization*, London and Sterling: Earthscan.

Wampler, B. and Avritzer, L. 2004. Participatory Publics: Civil Society and New Institutions in Democratic Brazil, *Comparative Politics* 36(3): 291–312.

Wapner, P. 1996. *Environmental Activism and World Civic Politics*, New York: State University of New York Press.

Weiss, T.G. 2012. *Thinking about Global Governance: Why People and Ideas Matter*, Abingdon: Routledge.

Whiddon, K. (forth). Sovereignty 'from Above?': Interrogating the Impacts of Global Governance on Natural Resource Tenure in Nepal, PhD dissertation, Centre for Agroecology, Water and Resilience, Coventry University, UK.

Wieviorka, M. 2005. After New Social Movements, *Social Movement Studies*, 1, May.

Wieviorka, M. 2012. The Resurgence of Social Movements, *Journal of Conflictology*, 3(2): 13–19.

Wieviorka, M. 2013. *Penser global. Socio*, 1, Mars. Lectures [En ligne], Les comptes rendus, 2013. http://lectures.revues.org/11469 (Accessed 12 February 2013).

Wieviorka, M. 2014. Social Movements, a Global Perspective, Keynote at the ISA World Congress of Sociology in Yokohama 9 July 2014. http://wieviorka.hypotheses.org/325 (Accessed 4 March 2015).

Wieviorka, M. and Calhoun, C. 2013. Manifeste pour les sciences sociales, *Socio*, 1: 3–38.

Willets, P. 2006. The Cardoso Report on the UN and Civil Society: Functionalism, Global Corporatism or Global Democracy? *Journal of Global Governance*, 12: 305–324.

Wittman, H.A. 2009. Reworking the Metabolic Rift: La Vía Campesina, Agrarian Citizenship, and Food Sovereignty, *Journal of Peasant Studies* 36(4): 805–826.

Wittman, H.A. 2012. Feeding the Nation while Mobilizing the Planet? La Vía Campesina, Food Sovereignty, and Agrarian Citizenship in Brazil. Paper for presentation at the Canadian Political Science Association 2012 Workshop: Feminist Political Ecology, Social Movements and Alternative Economy.

Wolford, W. 2010. Participatory Democracy by Default: Land Reform, Social Movements and the State in Northeastern Brazil, *Journal of Peasant Studies*, 37(1): 91–109.

World Bank. 2013. *Global Food Crisis Response Program*, 11 April 2013.

World Food Summit. 1996. Rome Declaration on World Food Security. World Food Summit 13. 17 November. http://www.fao.org/docrep/003/w3613e/w3613e00.htm (Accessed 19 February 2015).

Yang, G. 2000. Achieving Emotions in Collective Action: Emotional Processes and Movement Mobilization in the 1989 Chinese Student Movement, *Sociological Quarterly*, 41(4): 593–614.

Index

accountability, 7, 14, 35, 43, 44, 50, 116, 117, 121, 122, 133, 134, 140, 155, 157, 158, 166, 170, 186, 187, 192
activist researchers, 23
advocacy work, 11, 46, 58, 73, 120, 125, 128, 148, 150, 151, 158, 165, 168
agency, 4, 12, 13, 16, 17, 18, 19, 20, 21, 22, 23, 24, 25, 26, 60, 83, 95, 100, 102, 103, 104, 105, 108, 112, 113, 118, 147, 168, 169, 171, 176
agrarian activism, 14, 168
agrarian movements, 23, 80, 116, 143, 159
agribusiness model, 80, 93, 136
agroecology, 2, 30, 31, 63, 69, 72, 74, 76, 77, 89, 94, 152, 188, 190
alliance building, 11, 107
ambiguity, 147, 149
autonomous space, 44, 46, 56, 57
autonomy, 12, 13, 14, 20, 22, 30, 33, 35, 36–7, 41, 43, 46, 56, 57, 61, 62, 64, 99, 140, 142, 143, 148, 161, 163, 164, 170, 174, 185

boomerang model, 11, 83
bureaucratisation, 27, 39, 42, 132, 134, 141

capitalist modernity, 80
citizenship, 8, 90, 165
collective hermeneutic, 160
communication technologies, 40–1, 122
community building, 11
complex multilateralism, 172
compromise, 62, 70, 88, 91, 106
consensus, 37, 38, 54, 55, 56, 84, 86, 88, 94, 97, 99, 100, 105, 106, 114, 134
consultation, 5, 27, 35, 36, 44, 61, 66, 91, 92, 99, 116, 117, 133, 134, 140, 156, 172
contestation, 15, 78, 170
convergence, 8, 160
co-production of knowledge, 23, 159
cosmopolitan elite, 135, 140
cross-fertilisation, 157

damage control, 75
democratic control, 89, 90
democratic leadership, 116, 118, 120, 131, 141, 164
democratic spaces, 6, 150
de-politicisation, 76
de-radicalisation, 3, 13, 115, 140
discourse, 15, 32, 69, 72, 78, 99, 108, 117, 133, 134, 147, 148, 152, 178
diversity, 158

effectiveness, 82, 92, 100, 101, 131, 154
empowerment, 72, 103, 187, 188
entrusted leadership, 40, 123
entrusted researcher, 20, 22
exchange of experiences, 102, 156
exclusion, 13, 25, 36, 64, 66, 97, 98, 113, 150

food security, 4, 5, 13, 30, 31, 42, 49, 50, 61, 64, 67, 71, 72, 84, 85, 87, 88, 92, 93, 145, 186, 187, 188
food sovereignty, 7, 10, 13, 27, 29, 30, 31, 32, 45, 47, 48, 65, 66, 71, 74, 76, 81, 88, 90, 92, 93, 94, 97, 100, 107, 114, 118, 134, 136, 144, 151, 152, 158, 159, 160, 167, 168, 171, 175, 176, 178, 184, 185, 186, 188, 189, 190, 191, 192, 193
Freire, Paulo, 159

gender equality, 133, 187
global age society, 14, 169, 176, 177

global capitalism, 27
global civil society, 27, 35, 46, 53, 155, 157, 166
global decision-making, 83, 97, 174
global food governance, 4, 14, 89, 172, 174, 176
global governance, 6, 7, 8, 14, 34, 47, 82, 90, 117, 121, 169, 172, 177, 178
global policy agenda, 8, 164
global Strategic Framework, 64
globality, 170
globalization, 10, 11, 118, 166
GMO, 144, 192
grassroots activists, 118, 136

hegemonic meanings, 23
hierarchical organisational structures, 43
High Level Panel of Experts, 51, 81, 178
historicity of society, 18
human rights-based framework, 103

inclusion, 12, 13, 30, 35, 43, 37, 57, 76, 91, 97, 98, 99, 100, 113, 121, 131, 132, 133, 134, 150, 156, 169, 172, 173
independence, 143
indigenous peoples, 48, 50, 51, 60, 70, 71, 80, 94, 96, 98, 115, 176, 177, 184, 185, 187, 192
inequality, 25, 30
insider, 24, 26, 137, 138
interconnectivity, 116
internal democracy, 43, 99, 116, 117, 118, 119, 140
International Conference on Agrarian Reform and Rural Development, 48
International Federation of Agricultural Producers, 6, 34
International Food Security and Nutrition Civil Society Mechanism, 5, 12, 46
International Fund for Agricultural Development, 25, 45, 66, 141, 144, 166
International Panel of Experts on Sustainable Food Systems, 179
international peasant movement, 6, 27–45, 189
International Planning Committee for Food Sovereignty, 31, 46, 48, 93
interpreters, 2, 14, 95, 97, 98, 99, 104, 105, 114
investments in agriculture, 42, 61, 66, 70, 87, 99
iron cage, 141

Jakarta conference, 2, 20, 38, 45

knowledge production, 22, 23, 81, 169, 178, 179

land grabs, 2, 67, 137, 190
Landless Workers' Movement, 126
language, 13, 20, 21, 60, 78, 85, 95, 96, 97, 98, 99, 101, 104, 105, 113, 121, 124, 162, 175, 184
larger-scale transformation, 11
learning, 21, 23, 38, 60, 61, 83, 103, 110, 122, 123, 132, 156, 157, 158, 164, 170
legitimacy, 13, 43, 47, 53, 65, 78, 79, 84, 85, 90, 108, 116, 118, 123, 153, 154, 157, 158, 166, 167, 172
local communities, 6, 66, 69, 70, 175
local struggles, 10, 56, 120, 122, 126, 172, 175, 178, 189

Maputo Declaration, 29
meaning-making, 13, 15, 23, 65, 92
mistica, 96
mistrust, 107, 108, 148
modes of organisation/operation, 42–3
mother earth, 98
Movement of People Affected by Dams, 126
Movimento dos Pequenos Agricultores, 161, 167, 182
multi-actor governance, 79
multi-layered actor, 17
multi-scale activities, 4, 14, 124, 168, 169, 171, 172, 176
multi-scale and multi-arena research, 16

national food policies, 151
negotiation, 12, 18, 37, 53, 60, 69, 74, 86, 87, 93, 94, 95, 99, 104, 112, 138, 148, 149, 150, 173, 176, 188
network, 2, 9, 14, 19, 35, 39–42, 44, 48, 49, 63, 67, 82, 87, 93, 105, 109, 126, 130, 140, 182
Network of West African Peasant and Agricultural Producers' Organizations, 20
non-governmental organisations, 34, 35, 49
North–South divide, 27

one foot in, one foot out, 148
outsider, 24, 137, 138

participation, 5, 7, 12, 13, 14, 15, 18, 19, 22, 27, 32, 35, 37–9, 42, 43, 44, 46, 48, 51, 52, 53, 57, 58, 61, 62, 65, 67, 71, 76, 77, 82, 83, 85, 87, 90, 91, 95, 96, 97, 98, 100, 102, 104, 105, 109, 111, 113, 114, 115, 120, 121, 122, 124, 130, 131, 132, 133, 146, 150, 153, 154, 156, 157, 158, 161, 165, 166, 169, 173, 174, 177, 178, 186, 187
participatory methods, 43
passionate politics, 139
peasant identity, 12, 32, 33, 34, 35, 78, 97
peoples' struggles, 9, 81, 121, 125, 168, 174, 175, 178
political battlefield, 34, 65, 66, 67, 68, 69, 70, 71, 72, 73, 74, 75, 76, 77, 78, 79, 80, 81, 82, 83, 84, 85, 86, 87, 88, 89, 90, 91, 92, 93, 94, 117
political culture, 12, 33, 37, 44, 91, 133, 134, 140, 150, 164
political influence, 15, 55, 63, 77
political manoeuvring, 73
politicisation, 76
prefigurative activism, 118
principles for responsible agricultural investment that respects rights, livelihoods and resources, 186
Private Sector Mechanism, 53
protest, 3, 13, 32, 41, 61, 89, 149

small-scale farming, 176
social movements, 2, 3, 4, 10, 11, 12, 13, 14, 15, 17, 18, 19, 21, 22, 23, 24, 25, 26, 27, 35, 39, 41, 44, 48, 51, 52, 53, 56, 57, 58, 59, 62, 65, 71, 73, 74, 75, 76, 77, 78, 79, 80, 81, 82, 83, 87, 88, 91, 92, 93, 95, 96, 97, 98, 99, 100, 101, 102, 103, 105, 107, 112, 113, 114, 115, 116, 119, 122, 125, 126, 128, 130, 131, 134, 135, 136, 137, 138, 139, 140, 141, 144, 145, 146, 147, 150, 151, 152, 154, 155, 156, 157, 158, 159, 160, 161, 162, 163, 164, 166, 169, 170, 171, 172, 173, 174, 176, 177, 178, 179, 180, 182, 189, 193
solidarity research, 23
spaces of dialogues, 38
state bureaucracy, 95, 135
structure, 3, 4, 5, 11, 13, 17, 18, 38, 39–42, 43, 49, 95, 99, 113, 114, 116, 133, 135, 140, 141, 165, 179
subjectivity, 10, 11, 64, 78

technical support, 58, 64, 74, 105, 109, 114, 126, 133, 159, 183
top-down policies, 35
top-down structure, 12, 27, 43
transdisciplinarity research, 179
translators, 8, 104, 105, 121
transnational activism, 11, 156, 171
transnational networks, 8, 11, 83
transnational policy-making, 65, 92, 113, 174
trust, 12, 19, 20, 24, 27, 36, 40, 59, 62, 107, 108, 141, 144, 157

United Nations Commission on Human Rights Council, 31
United Nations Conference on Trade and Development, 145
United Nations Declaration on the Rights of Peasants and Other People Working in Rural Areas, 145
United Nations General Assembly, 49
United Nations language, 97, 98

voice, 1, 3, 6, 8, 13, 21, 27, 29, 34, 35, 53, 54, 55, 74, 87, 88, 90, 116, 119, 146, 168, 169, 174, 177, 178
vote, 55, 56, 74, 88, 91, 144

way of reason, 11
World Bank, 9, 66, 113, 143, 144, 155, 165, 186
World Food Summit, 29, 46, 47, 48
World Social Forum, 35, 67, 68, 90, 118, 130, 166, 182
World Trade Organization, 7, 8, 9, 49